Digital Technologies and Tools for Smart Agriculture

To meet the food requirements of the growing population worldwide, the agricultural domain is being technologically empowered to produce more with fewer resources and investments. Fortunately, the faster maturity and stability of digital (digitization and digitalization) technologies and tools have laid a stimulating and sparking foundation for large-scale production of food items with minimal time, talent, and treasure. The penetration and pervasiveness of the Internet of Things (IoT) sensors and actuators are considered the first and foremost aspect of advanced and adaptive agriculture. Further, the device ecosystem for automated and accelerated agriculture processes and practices grows consistently. The robots and drone space advancements are fascinating, as they are set to transform the industry. Connected robots, drones, and other agricultural equipment, appliances, instruments, and machinery are flourishing towards intelligent agriculture. Agricultural sensors and devices generate a lot of real-time and real-world digital data. With the help of artificial intelligence (AI) methods, all agriculture data are subjected to various deep and decisive investigations to extract actionable insights in time. Other noteworthy improvisations include the widespread deployment of 5G communication networks, the growing power of edge analytics and intelligence, the distinct contributions of blockchain technology, the adoption and adaptation of cloud-native computing principles, the phenomenon of digital twins, etc., in the digital space. All these developments have led to the realization of smart agriculture products, solutions, and services.

Features:

- Delineate the role and responsibilities of digital technologies towards smart agriculture
- Delve into the potential of IoT and AI, along with other critical technological advancement
- Demystify large language models (LLMs) and visual language models (VLMs)
- Dig deep and delve into the advancements in the robots and drone space
- Describe how digital innovations can significantly elevate agriculture processes, platforms, patterns, and practices

This book will delve into the transformative power of digital technologies in agriculture, particularly in the context of smart agriculture. It will highlight how these technologies and innovative digital devices are reshaping the agricultural landscape. The book will also showcase the disruptive potential of these technologies, illustrating how they are revolutionizing traditional farming practices and processes.

Digital Technologies and Tools for Smart Agriculture

Edited by
Madhumathi Ramasamy, Karthigha Mohan,
Pethuru Raj Chelliah, and Kai Sheng

CRC Press
Taylor & Francis Group
Boca Raton London New York

CRC Press is an imprint of the
Taylor & Francis Group, an **informa** business

Designed cover image: Shutterstock

First edition published 2025
by CRC Press
2385 NW Executive Center Drive, Suite 320, Boca Raton FL 33431

and by CRC Press
4 Park Square, Milton Park, Abingdon, Oxon, OX14 4RN

CRC Press is an imprint of Taylor & Francis Group, LLC

ISBN: 9781032782553 (hbk)
ISBN: 9781032782560 (pbk)
ISBN: 9781003487005 (ebk)

DOI: 10.1201/9781003487005

Typeset in Times
by Newgen Publishing UK

Contents

Preface

A dazzling array of digital technologies is sophisticatedly empowering several business domains in this digital era and inspiring a new wave of innovation and growth. This distinct saga starts with the blossoming aspect of digitization. All physical, mechanical, and electrical systems are being methodically digitized by digitization and edge technologies. Further on, digitized entities are being succulently enabled through new network topologies and communication technologies. Such an enablement helps digital devices join mainstream computing. They also can communicate and collaborate on a need basis to accomplish specialized and localized tasks with all clarity and alacrity. Further on, these devices at the edge can interact with cloud-hosted software solutions (on and off-premises) to be artistically and adequately empowered to guarantee versatile jobs resiliently. One noteworthy transition with the splurge of these IoT devices across our everyday environments (hotels, hospitals, homes, etc.) is that context information gets meticulously collected and shared in time to envisage real-time and real-world software services and applications.

Edge and digitization technologies such as sensors, actuators, stickers, RFID tags, barcodes, microcontrollers, LED lights, beacons, etc., transition any ordinary entities in our midst into digitized elements. Digitized entities, electronics (consumer and industry), business workloads, and IT systems synchronize flawlessly to create more intelligent devices, systems, networks, and environments. It is predicted that there will be millions of software services (microservices), billions of connected devices, and trillions of digital entities on Earth. When these assets and artifacts interact purposefully, a massive quantity of multi-structured data will exist. With the faster maturity and stability of artificial intelligence (AI) algorithms, models, agents, frameworks, and accelerators, digital data generated and collected is subjected to a variety of more profound and decisive investigations through the intelligent leverage of AI algorithms (machine and deep learning (ML/DL) algorithms) to extract actionable insights. This technology-powered knowledge discovery out of data volumes has set a stimulating and sparkling foundation for digital intelligence, empowering decision-making with informed and actionable insights.

This book is a comprehensive exploration of the role of digital technologies in agriculture. We will delve into the potential of IoT and AI, along with other key technological advancements such as 5G, edge computing, blockchain, digital twins, edge AI, large language models (LLMs), robots, and drones. We aim to visualize how these innovations can significantly elevate agriculture processes, platforms, patterns, and practices, ultimately reducing the fear of starvation.

About the Editors

Madhumathi Ramasamy completed her PhD from Anna University, Chennai in November 2016. She has around 22 years of teaching experience including 12 years of research experience. Her area of interest includes cloud computing, Artificial Intelligence, Internet of Things, and quantum computing. She has published many papers in SCI International Journals and IEEE, Springer conferences. She has received research grants from the Department of Science and Technology, Science and Engineering Research Board (DST-SERB) under Teachers Associateship for Research Excellence (TARE) scheme. Recently she has received a visiting scientist's fellowship from the Indian National Science and Academy (INSA) to work in IIT-Madras. She has also received grants from Indian Council for Medical Research (ICMR), Science and Engineering Research Board, and Council of Scientific and Industrial Research (CSIR) to conduct various workshops and seminars. She has chaired sessions for many National and International Conferences. She is a life member in CSI, ISTE, ISAE, and member in IEEE Women in Engineering.

Academic url: https://srec.ac.in/department/cse/faculty/details/161
LinkedIn: www.linkedin.com/in/madhumathi-sengottaiyan-4317708b/
Google scholar https://scholar.google.com/citations?user=648PmAwAAAAJ&hl=en

Karthigha Mohan is presently working as an Assistant Professor, Department of Computing Science and Engineering, PSG Institute of Technology and Applied Research, Coimbatore, India. She completed her PhD degree in the area of network security and deep learning. She has 12 years of experience which includes both teaching and research. She has been guiding undergraduates and postgraduates in all the recent areas of computing security. She has published more than 20 research papers in national and international journals, international conferences, and book chapters. She has delivered technical talks in the areas of cyber security and AR/VR. Her research areas include cyber security, deep learning, virtual reality, and augmented reality.

Academic url: https://cse.psgitech.ac.in/cse/faculty
LinkedIn: www.linkedin.com/in/karthigha-mohan-189027157/
Google Scholar: https://scholar.google.com/citations?user=abJqF8sAAAAJ&hl=en&authuser=1

Pethuru Raj Chelliah works as a chief architect at Edge AI Division, Reliance Jio Platforms Ltd. (JPL) Bangalore. Previously, worked in IBM global Cloud center of Excellence (CoE), Wipro consulting services (WCS), and Robert Bosch Corporate Research (CR). In total, I have gained more than 22 years of IT industry experience and 8 years of research experience. Finished the CSIR-sponsored PhD degree at Anna University, Chennai and continued with the UGC-sponsored post-doctoral research in the Department of Computer Science and Automation, Indian Institute of Science (IISc), Bangalore. Thereafter, I was granted a couple of international research fellowships (JSPS and JST) to work as a research scientist for 3.5 years in two leading Japanese universities. I am a professional member in ACM, focusing on some of the emerging technologies such as the Internet of Things (IoT), Artificial Intelligence (AI) Model engineering and optimization, big, fast and streaming analytics, blockchain, digital twins, cloud-native computing, edge and serverless computing, reliability engineering, Microservices architecture (MSA), event-driven architecture (EDA), 5G, etc. More about my academic and industry contributions is in

Technology Books: https://peterindia.net/MyBooks.html
A Portal on Digital Technologies: www.peterindia.net/Digital.html
Publications: https://scholar.google.co.in/citations?user=yaDflpYAAAAJ&hl=en
LinkedIn: www.linkedin.com/in/sweetypeter/

Kai Sheng received the BE degree in information engineering from the Central South University, Changsha, China, in 2000 and PhD degree in risk engineering from the University of Tsukuba, Tsukuba, Japan, in 2006. He has worked as an engineer at CARMATE (Japan) and Dongfeng Motor Corporation (China). He is currently a Professor of the Guangzhou Institute of Technology and the Academy of Advanced Interdisciplinary Research, Xidian University, China. His research interests include computational intelligence, remote sensing image understanding, and the V2X-based automated driving system development.

Contributors

Aaskaran Bishnoi
Department of AIT-CSE
Chandigarh University

Abhishek Gawali
Department of Information Technology
G. H. Raisoni College of Engineering
Nagpur

Arvind Kumar
Associate Professor, SCA
MRIIRS

Balaji Kannan
Professor (SWC) & Head, Department of Physical Sciences & IT
Agricultural Engineering College & Research Institute
Tamil Nadu Agricultural University

Dhanashree Diwase
Department of Information Technology
G. H. Raisoni College of Engineering
Nagpur

Ganga A
Department of Biotechnology
Kamaraj College of Engineering and Technology
Madurai

Ganga M
Associate Professor, Department of Biomedical Engineering
Hindusthan College of Engineering and Technology
Coimbatore, Tamil Nadu, India

Garima Singh
Department of Computer Science and Engineering
IIT (BHU) Varanasi

Gurwinder Singh
Department of AIT-CSE
Chandigarh University

Janani N
Research Scholar, Department of Soil Water Conservation Engineering
Agricultural Engineering College & Research Institute
Tamil Nadu Agricultural University

Kamaraj Arunachalam
Department of Electronics and Communication Engineering
Mepco Schlenk Engineering College
Sivakasi

Karthigha Mohan
Assistant Professor
Department of Computer Science and Engineering
PSG Institute of Technology and Applied Research

Karishma C G
Teaching Assistant
Agricultural Engineering College & Research Institute
Tamil Nadu Agricultural University

Kavino M
Professor (Horticulture)
Department of Fruit Science
Tamil Nadu Agricultural University
Coimbatore, Tamil Nadu, India

Kavitha S
Department of General Engineering
Kamaraj College of Engineering and Technology
Madurai

Kaviya P
Department of Information Technology
Kamaraj College of Engineering and Technology
Madurai

Krishna Pandey
Assistant Professor, Department of Journalism & Mass Communication
School of Media Studies & Humanities
Manav Rachna International Institute of Research & Studies
Faridabad

Madhumathi Ramasamy
Associate Professor
Department of Computer Science and Engineering
Sri Ramakrishna Engineering College

Mohammad Shahnawaz Shaikh
Department of AI & DS
Parul Institute of Engineering and Technology
Parul University
Vadodra, Gujarat, India

Nirmala Devi M
Assistant Professor (Mathematics)
Department of Physical Sciences and Information Technology
AEC & RI
Tamil Nadu Agricultural University
Coimbatore, Tamil Nadu, India

Prasanth Kumar Bevara
Department of Computer Science and Engineering
IIT (BHU) Varanasi

Preethi S
Sri Ramakrishna Engineering College
Coimbatore, Tamil Nadu

Rahul Joshi
Assistant Professor, Department of Journalism & Mass Communication
School of Media Studies & Humanities
Manav Rachna International Institute of Research & Studies
Faridabad

Ramanan M
Teaching Assistant
Agricultural Engineering College & Research Institute
Tamil Nadu Agricultural University

Ranjit Singh
Department of AIT-CSE
Chandigarh University

Ranjitham G
Associate Professor
Department of Electronics and Communication Engineering
Sri Krishna College of Engineering and Technology
Coimbatore, Tamil Nadu, India

Ravi Shankar Singh
Department of Computer Science and Engineering
IIT (BHU) Varanasi

Ravindrakumar Selvaraj
Professor
Biomedical Engineering
Kalaignarkarunanidhi Institute of Technology
Coimbatore, Tamil Nadu, India

Romir Mehta
Department of Computer Science and Engineering
IIT (BHU) Varanasi

Ronalda M
Department of Computer Science and Engineering
National Institute of Technology
Srinagar, Jammu and Kashmir, India

Rupali Vairagade
Department of Cyber Security,
Shah and Anchor Kutchhi Engineering College, Mumbai

Sakthi P
Assistant Professor, Department of Electronics and Communication Engineering
M. Kumarasamy College of Engineering

Sangeetha
Department of CSE
SRM Institute of Science and Technology
Chennai

Saravana Sundaram S
Professor and Head
Department of Biomedical Engineering
Hindusthan College of Engineering and Technology
Coimbatore, Tamil Nadu, India

Sathish Kumar Nagarajan
Professor
Department of Electronics and Communication Engineering
Sri Ramakrishna Engineering College
Coimbatore, Tamil Nadu, India

Sathya S
Sri Ramakrishna Engineering College,
Coimbatore, Tamil Nadu

Selvakumar B
Department of Computer Science and Engineering
Mepco Schlenk Engineering College
Sivakasi, Tamil Nadu, India

Selvan C
Department of Computer Science and Engineering
Reva University
Yelahanka, Kattigenahalli, Bengaluru, Sathanur, Karnataka, India

Senthil Kumar Jagatheesaperumal
Department of Electronics and Communication Engineering
Mepco Schlenk Engineering College
Sivakasi

Shruthi Ravikumar
Sri Ramakrishna Engineering College,
Coimbatore, Tamil Nadu

Shweta Kushwaha
Department of Computer Science and Engineering
IIT (BHU) Varanasi

Simrat Waila
Department of Computer Science Engineering
Baba Banda Singh Bahadur Engineering College
Fatehgarh Sahib

Sneha Iyer
Sri Ramakrishna Engineering College
Coimbatore, Tamil Nadu

Sripriyan K
Assistant Professor
School of Mechanical Engineering
VIT-Bhopal

Stuti Tandon
Assistant Professor, SCA
MRIIRS

Sucharitha M
Associate Professor, School of Electronics Engineering
VIT-AP University

Suhail Javed Quraishi
SCA
MRIIRS

Sumathi C S
Associate Professor (CS), Department of Physical Sciences & IT
Agricultural Engineering College & Research Institute
Tamil Nadu Agricultural University

Suman Kumari
Associate Professor, Department of Journalism & Mass Communication
School of Media Studies & Humanities
Manav Rachna International Institute of Research & Studies
Faridabad

Uma J
Professor, Department of Electrical and Electronics Engineering
M. Kumarasamy College of Engineering

Veeramani P
Assistant Professor
Department of Electronics and Instrumentation Engineering
Sri Ramakrishna Engineering College
Coimbatore, Tamil Nadu, India

Veningston K
Department of Computer Science and Engineering
National Institute of Technology
Srinagar

Vicky Kumar Nayak
Department of Computer Science and Engineering
IIT (BHU) Varanasi

1 New Generation AI Algorithms and Models for Smart Agriculture

M. Kavino, M. Ganga, and S. Saravana Sundaram

1.1 INTRODUCTION

It is anticipated that by 2050, there will be nearly10 billion people on the planet, putting extreme strain on the agricultural industry to boost yields of crops and productivity. Agriculture in the current era is changing and taking many creative turns. Faced with a number of challenges in reaching the targeted levels of agricultural productivity, including dwindling land resources and soil fertility, scarcity in labour, climate change, small land holdings, and environmental problems. In order to mitigate the looming food scarcity, two possible strategies have surfaced: increasing land use and implementing large-scale farming or embracing creative methods and utilizing technology to boost output on currently cultivated land. The days of manual ploughs and horse-drawn machinery are definitely behind us in terms of farming technology.

The introduction of new technology is intended to maximize yield and boost productivity. Both small-scale farmers and huge worldwide agribusinesses, however, frequently overlook the opportunities that artificial intelligence in agriculture might bring to their farming operations.

1.2 ARTIFICIAL INTELLIGENCE (AI)

Artificial intelligence is an interdisciplinary branch of study that models human actions and cognition, including learning and problem-solving, in an effort to replicate human intelligence in robots. AI technology is presently being used by researchers and extension agents to address issues related to agricultural productivity. By aiding in the selection of appropriate crop varieties, the adoption of improved fertilizer and soil management techniques, the control of pests and diseases, the computation of agricultural productivity and the forecasting of commodity prices, artificial intelligence can assist farmers in increasing yields.

In order to tackle agricultural issues, artificial intelligence uses a range of state-of-the-art methods, including deep learning, robots, the Internet of Things, artificial neural networks, wireless sensor networks, image processing, and machine learning [1] & [2].

DOI: 10.1201/9781003487005-1

1.3 ARTIFICIAL INTELLIGENCE MODEL

Among the most fascinating areas in computer science is AI. The goal of AI research is to construct a machine that can instantly mimic human intelligence. It is still impossible to create true AI tools with human-like intelligence. This does not negate the potential benefits of applying AI algorithms, though. These days, AI models are available for various analytical and judgemental activities. An AI model is a program or algorithm that uses training data to spot trends and generate predictions or judgement calls. An AI model's ability to analyze and forecast data more accurately increases with the quantity of data points it receives. In addition, AI models employ decision-making algorithms to gather and analyze data, apply what they have learned to accomplish predetermined objectives, and learn from their training. AI models excel at using a lot of data to solve complicated problems. They are therefore able to solve complicated problems with a great degree of accuracy.

1.4 RECENTLY DEVELOPED AI MODELS

1.4.1 YOLOv8

One of the most popular real-time machine vision algorithms is YOLO (You Only Look Once), which was first introduced by Redmon [3]. Using characteristics acquired from a deep convolutional neural network, it quickly and accurately identifies specific objects in photos, videos or live feeds while functioning at a considerably quicker rate than other networks. The most advanced YOLO model, YOLOv8, was just launched and has already set new benchmarks for real-time recognition and segmentation. It is expected to become a popular topic in study because it uses AI to solve complicated problems in a straightforward and efficient manner.

Object detection is approached as a simple regression issue by single-stage detectors such as You Only Look Once (YOLO) and Single Shot MultiBox Detector (SSD), which take an input image and learn the bounding box coordinates and class probabilities. Conversely, these models achieve poorer accuracy but are faster. These algorithms have been used by researchers in agriculture over time to recognize fruits, vegetables and other agricultural goods. To a certain extent, similar algorithms have also been used to determine the growth phases of these kinds of produce. Furthermore, these methods have been used sparingly in the tracking and counting sector. Tian et al. [4] demonstrated an improved YOLOv3 model for apple identification in orchards with varied lighting, complex backgrounds, overlapping apples and branches and leaves at different stages of growth. The model had an average detection time of 0.304 s per frame and an F1 score of 0.809 for ripe apples, according to the test findings. Liu and Wang [5] enhanced the YOlOv3 model's feature layer to improve speed and accuracy of detection after creating a real-world dataset of pests and illnesses affecting tomatoes. According to test results, this permitted a mean average precision (mAP) of 92.39% and a mean detection time of 20.39 milliseconds.

Wu et al. [6] presented a real-time apple flower detection method based on the channel-pruned YOLOv4 deep learning algorithm. The test findings showed that the pruned apple blossom identification model had a 97.31% mAP, a 231.51 MB decrease in model size, a 39.47 % drop in inference time and a 96.74% reduction in parameters—all of which were just 0.24% lower than the model before pruning. Green and reddish tomatoes were distinguished by Magalhães et al. [7] using the Single-Shot MultiBox Detector (SSD) and YOLO designs. They found that SSD MobileNet v2 outperformed SSD Inception v2, SSD ResNet 50, SSD ResNet 101, and YOLOv4 Tiny, with an F1-score of 66.15%, a mAP of 51.46%, and an inference time of 16.44 ms. Bazame et al. [8] carried out coffee berry detection and classification using the YOLOv3-tiny. For the classes of unripe, ripe and overripe coffee berry, the average precision was 86%, 85%, and 80%, respectively. For the validation set, the findings showed a mAP of 84%, an F1-Score of 82%, precision of 83%, and recall of 82%. to identify 'Abbe Fetetl' pear ripening stage and physiological problems. An et al. [9] developed an algorithm named SDNet (Strawberry Detect Net) for detecting strawberry growth. According to the testing results, SDNet has a monitoring speed of 30.5 ms and a precision, accuracy and recall of 94.26%, 93.15% and 90.72%, respectively. Using a for-profit online platform, Bonora et al. [10] evaluated convolution neural networks with "You Only Look Once" (YOLO) architecture.

Gai et al. [11] proposed an improved YOLOv4 deep learning algorithm to identify cherry fruits with a compact volume. They also showed that the mAP (average accuracy) value obtained by using the improved YOLOv4 model (YOLOv4-dense) network was 0.15 higher than that of the state-of-the-art YOLOv4. Li et al. [12] presented a fast and lightweight detection technique based on YOLOv5 to boost the recognition rate and lower the detection time for pests related to passion fruit. A lightweight tea bud identification model based on the YOLOv5 model was reported by Gui et al. [13], who also demonstrated that the modified model's mean average precision increased by 9.66% in comparison to the YOLOv5 model. Liu et al. [14] proposed a method for the perception of soybean phenotypic information based on improved YOLOv5.

Paul et al. [15] introduced the use of YOLO segmentation models for peduncle detection and a variety of YOLO methods for the detection of capsicums. YOLOv8s met the same IOU threshold and obtained a respectable mAP of 0.614 in the development stage determination challenge. The YOLOv8s-seg model also performed better in peduncle recognition, achieving mask and box mAPs of 0.771 and 0.790, respectively [15].

Shi et al. [16] offered a combined multi-crop row centreline extraction algorithm. This algorithm is based on the B-spline curve, least squares approach, threshold DBSCAN (Density-Based Spatial Clustering of Applications with Noise) clustering, and improved YOLOv8 (You Only Look Once-v8) model [17].

1.4.2 Segment Anything Model (SAM)

Segment Anything Model is a computer vision model for picture segmentation that was trained on a new dataset and released by Meta Research in April, 2023. Segment

Anything Model (SAM) is the name of the model itself. Segmentation masks for all objects in an image that the model can locate or for objects that match a given text prompt can be created using SAM. Eleven million photos and more than one billion image masks make up the dataset used to train SAM.

Of late, segmentation foundation models have seen tremendous advancements in natural image segmentation [18], which allows for accurate and effective object segmentation in a fully automatic or interactive manner. Pre-trained weights are used in these models, which are usually based on transformer topologies, to deliver modern efficiency and unparalleled generalization ability on a variety of realistic images. One of these is a picture segmentation challenge, model and dataset called the Segment Anything project [19] from Meta AI Research Meta (Menlo Park, CA, USA). The largest segmentation dataset to date, SA-1B [20], which contains over 1 billion masks on 11 million pictures, was used to train SAM. In order to extend and adapt its segmentation capabilities to new picture distributions and jobs, SAM is trained and intended to be promptable.

Segment Anything Model (SAM) was introduced by Carraro et al [21] with the goal of speeding up the revolution in smart farming. Many precision agriculture challenges require the use of computer vision techniques implemented by supervised deep learning systems, which requires extensive data gathering and time-consuming human ground truth annotation. In artificial intelligence (AI), the so-called foundation models are becoming more and more important. In order to provide a foundational model for image segmentation, Meta AI Research is working on a project named Segment Anything. Without further training, it can achieve zero-shot generalization to unfamiliar objects and images. In the perspective of precision, this study assesses the Segment Anything Model's (SAM) performance in solving the semantic segmentation of objects problem [21].

1.5 BENEFITS OF AI IN AGRICULTURE

Agriculture has been the foundation of human civilization for thousands of years, providing both food and economic expansion. Contrarily, artificial intelligence (AI) has only been around for a few decades, even in its most rudimentary forms. But new ideas are being introduced in every industry and agriculture is no exception. Recent years have seen a rapid advancement in agricultural technology, which has revolutionized farming methods. Global problems like population growth, climate change and resource scarcity are posing a danger to the sustainability of our food system, which is why these technologies are becoming more and more crucial. Numerous issues are solved and the drawbacks of conventional farming are reduced with the introduction of AI [22].

The rate of urbanization is rising in tandem with global population growth. Consumer behaviour is evolving due to rising disposable income. Farmers must devise a plan to increase output because they are under extreme pressure to satisfy the growing demand of food grains. A farmer must be creative because there is a limited availability of fertile land. Plans must be developed to assist farmers in managing their risks or, at the absolute least, lowering them. The potential for worldwide adoption of AI in agriculture is one of the most exciting opportunities.

These risks are increased by monoculture, widespread pesticide usage, and climate change. When combined, these elements provide farmers a new challenge. Farms and farmers face a tremendous degree of stress because most products produced on farms are mostly dependent on natural sources. This stress is exacerbated by the need for increased yields each year, weather fluctuations, and a labour shortage. This suggests that farm efficiency would need to double and the agricultural sector will need to develop dramatically in the coming years if we are to even remotely accomplish our targets. AI provides automation for agriculture while accounting for all of these challenges.

1.6 APPLICATIONS OF ARTIFICIAL INTELLIGENCE IN AGRICULTURE

Similar to traditional agricultural operations, farmers face a multitude of challenges. In order to solve these problems, AI is being applied widely in this industry. Agriculture has found artificial intelligence to be a game-changer in the recent past. It helps farmers in a variety of ways, including healthier crop production, insect and soil detection and control and soil management. Some noteworthy applications of AI in agriculture (Figure 1.1) are as follows:

1.6.1 Weather and Price Forecasting

Climate change makes farming a challenge for farmers to make the best decisions on soil preparation, sowing and harvesting. However, farmers can prepare for the type of crop to produce, the seeds to sow and the crop harvesting process with the use of AI weather forecasting, which provides information on weather analysis. Price forecasting helps farmers make the most profit possible by giving them a clearer estimate of the price of crops over the following few weeks.

FIGURE 1.1 Applications of artificial intelligence in agriculture.

1.6.2 Health Monitoring of Crops

The kind and nutrition of the soil have a significant impact on crop production and quality of harvestable products. However, the quality of the soil is deteriorating due to the increasing pace of deforestation and this is difficult to measure.

Artificial intelligence has developed a new program called Plantix to address this problem. Plantix Ecosystem and Agro Technology (PEAT) created it to detect soil inadequacies, such as illnesses and pests that affect plants. Farmers can use this application to gain ideas on how to apply better fertilizer, which will raise the quality of the harvest. Through the use of AI's image recognition technology in this app, farmers may take pictures of their plants and obtain quality information.

1.6.3 Agriculture Robotics

Many industries, namely the manufacturing sector, use robotics extensively to carry out difficult jobs. Currently, a number of AI startups are creating robots for use in the agricultural industry. These AI robots are designed to be capable of carrying out various farming duties. AI robots are also trained to inspect the quality of harvested produce, identify and manage weeds and harvest crops more quickly than a human.

1.6.4 Intelligent Spraying

Weed is easily recognized by AI sensors, which can also identify locations where weed growth has occurred. Herbicides can be accurately applied in these locations to minimize their use and save time and crop damage. Numerous artificial intelligence companies are developing robots that use artificial intelligence and computer vision to precisely spray herbicides. AI sprayers have the potential to significantly reduce the amount of chemicals used in fields, improving crop quality and resulting in cost savings.

1.6.5 Disease Diagnosis

With AI forecasts, farmers can simply learn about illnesses. With this, they can quickly and accurately detect illnesses using the right approach. It can spare farmers' time and save the lives of plants. First, computer vision technology is used to preprocess plant photos in order to do this. This guarantees that plant photos are correctly segmented among sections that are unhealthy and sections that are not. The diseased portion is cut after it is discovered and sent to the lab for additional diagnosis. In addition, this method aids in the identification of pests, nutrient deficiencies and many other issues.

1.6.6. Precision Farming

It is all about "Right place, Right time and Right products" in precision farming. The labour-intensive portion of farming that involves carrying out repetitive chores can be replaced with a far more precise and regulated method called precision farming. Plant stress level identification is one application of precision farming. High-resolution photos and various plant sensor data can be used to obtain this. After that, a machine learning model receives the sensor data as input to identify stress.

In order to maximize profit, farmers can assess crop yields and forecast the ideal time for harvesting based on the maturity indices using artificial intelligence and soil health in real-time. The early identification of plant diseases will also enable farmers to employ the most effective control measure in appropriate time with label specific fungicides. It is feasible to harvest fruits on behalf of workers with greater precision and speed by using sensors, machine vision, AI models and robots. Additionally, it aids in minimizing crop wastage in the field that is brought about by the conventional way of harvesting like shaking tree branches which leads to falling of mature and immature fruits. It is feasible to forecast the ideal moment to fertilize fields based on the soil and crop leaf nutrient status and sow seeds using artificial intelligence techniques and tools which pave the way for smart horticulture, resulting in a higher yield and better prices at the right time. Label specific pesticides and fungicides spraying are regarded as a crucial management technique for pest and disease control in the context of sustainable agriculture.

1.7 ARTIFICIAL INTELLIGENCE APPLICATIONS IN HORTICULTURE

Prediction of crop yield: It is feasible to forecast the ideal moment to fertilize crops based on the available soil and plant leaf nutrient status and sow seeds using artificial intelligence techniques and tools, resulting in a higher yield with low inputs usage and better prices at the right time.

Monitoring of soil and crops: AI and machine learning, for instance, can be used by farmers to estimate crop yields and forecast the most profitable time for harvesting by choosing right maturity indices and soil health.

Diagnosis of disease: Farmers will take decisions to apply the most effective fungicides or other disease management strategies with early identification of plant diseases.

Horticulture robot to do labor-intensive farm duties: Harvesting work can be completed faster and more accurately on behalf of workers by using robots, sensors, machine vision, AI models and AI systems. By effectively using AI tools, it is possible to reduce the post-harvest wastage considerably.

Strategic spraying: The large area under fruit orchards can be easily covered in a short time whenever we choose aerial spray using AI tools i.e. Drone spray.

1.8 APPLICATIONS OF ARTIFICIAL INTELLIGENCE IN FRUIT CROPS

Researchers have currently assembled a number of open-access image datasets featuring various horticultural crops, including ImageNet and Open University Fruits Dataset (OUFD). These deep neural network models are trained using large-scale image datasets, which are often employed and offer a wealth of feature information for horticultural crops recognition. Nevertheless, horticultural crop identification

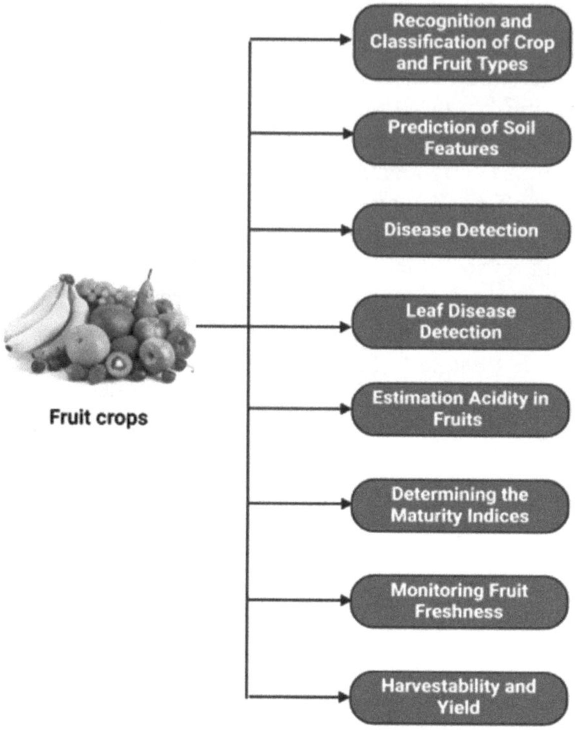

FIGURE 1.2 Applications of artificial intelligence in fruit crops.

algorithms are still in the early stages of development and have not yet been widely used, the majority of researchers preferring to conduct experiments using their own collection of image sets (Liu et al. [14]). The major application of artificial intelligence in fruit crops are mentioned in Figure 1.2.

1.8.1 Recognition and Classification of Crop and Fruit Types

Mandal et al. [23] implemented an SVM classifier for six labels: rice, sugarcane, cotton, bananas, bare fields and mixed classes. The predictive model used images of satellite multi-temporal crop classification. The image dataset described the geometric features of the crop from a 4-month interval covering the critical crop-growth stages.

In order to classify fruits, a hybrid deep learning approach has been suggested by Xue et al. [24]. It is inferred that the majority of studies use modified Faster-RCNN and Masked CNN in addition to modified ResNet or VGG as feature extractors. The YOLO (You Only Look Once) has gained prominence in the field of object identification at the moment because of its capacity to identify minute objects, with different versions being used by different researchers to identify fruits. Varietal discrimination of guava (*Psidium guajava*) leaves using multi-features analysis is described in a paper written by Asim et al. [25]. This study looked at the potential of Machine Vision (MV) methods for categorizing and identifying 12 different guava types. To distinguish between various guava plant species, Mostafa et al. [26] suggested a deep convolutional neural network (DCNN)-based data improvement method employing color-histogram equalization and the unsharp masking technique. In order to enhance the quantity of altered plant images, nine angles from 360 were employed.

1.8.2 Prediction of Soil Attributes

The majority of soil attributes are expensive, time-consuming and subject to change. To overcome the paucity of measurable soil property information, quick and accurate soil property prediction is essential. Since the advent of geostatistics, which allowed for the calculation of forecast accuracy and error, significant advancements have been made in this area over the last 30 years.

An ML model was developed by Cevallos et al. [27] to evaluate soil quality (classifying it into three categories: ideal, acceptable and unacceptable). In the study, soil samples were taken from 0 to 30 cm of the banana crop's field to know the macro and micro nutrients status (N, P, K, Ca, Mg, Fe, Mn, Cu, and Zn). In order to compare the results of the seven ML algorithms and select the most accurate predictive model, the author utilized MLP, BN, LGR, DT, RF, SVM, and KNN.

The Crop Prediction Model Using Machine Learning Algorithms was created by Elbasi et al. [28]. These algorithms' primary goal is to maximize crop yield and minimize waste by making wise choices about when and how to sow, irrigate, and harvest crops. Elbasi et al. [28] has developed the Crop Prediction Model Using Machine Learning Algorithms. The main focus of these algorithms is to help optimize crop production and reduce waste through informed decisions regarding planting, watering, and harvesting crops.

1.8.3 Disease Detection

A vital part of human civilization, agricultural biodiversity is necessary to supply humans with food and raw materials (Tirkey et al. [29] and Ramanjot et al. [30]). Illness can develop when a plant is consistently harmed by pathogenic organisms such fungi, bacteria and nematodes; changes in soil pH, high or low temperatures, variations in the amount of moisture and humidity in the air, and other factors.

Yang et al. [31] developed models employing deep learning neural network models such as VGG16, Inception V3, ResNet50 and ResNet101 based on more than 7000 images of pear disease to investigate the relationship between important influencing elements and severity of pear disease. The diagnosis models' recognition rates ranged from 97.67% to 99.44%. An artificial intelligence (AI) driven framework is offered by Almadhor et al. [32] to identify and categorize the most prevalent guava plant diseases. The suggested framework uses 4E colour difference image segmentation to divide the diseased areas. To extract rich, useful feature vectors, colour (RGB, HSV) histogram and texture (LBP) characteristics are also used.

1.8.4 Leaf Disease Detection

A condition that arises naturally from a plant's growth, leaf disease not only causes obstacles in the agribusiness sector but also lowers a nation's agricultural output. In their life cycles, a variety of bacteria, fungi, viruses and other naturally occurring infectious organisms are the primary causes of leaf disease. Leaf stresses can be identified and categorized in a variety of ways. The first, less common method, is direct visual inspection with the naked eye. Second, leaf stresses can be examined using machine learning (ML) techniques or a manual procedure. Applying certain machine learning algorithms to plant leaves is the next better alternative in this comparable trend that is taken into consideration in many forms of research. Early on, it is difficult to identify leaf diseases before they spread by applying past information. According to Sun et al., [33], Apple Disease [18] is a dataset of images of apple leaves with five common leaf diseases: alternaria blotch, brown spot, grey spot, mosaic spot and rust. The experiments are compared to the most recent state-of-the-art CNN-based apple disease recognition approaches, including MEAN-SSD (Sun et al., [33]) and LWCNN [34].

1.8.5 Estimation of Acidity in Fruits

Fruits and vegetables' organoleptic qualities—specifically, their flavour, colour and aroma—are greatly influenced by organic acids. They are in charge of the acidity or sourness of fruit, which is measured in relation to citric acid [35]. Furthermore, organic acids function as precursors of phenolics and taste compounds and indirectly influence the metabolism of phenols by changing pH. Analyzing the organic acid profile is crucial for characterizing various cultivars and researching how agronomic conditions or maturity affect fruit quality. Numerous techniques have been devised to distinguish and measure specific organic acids in various vegetative matrixes.

Using metaheuristic algorithms, Pourdarbani [36] developed a model to assess acidity of Fuji apples. To estimate the acidity qualities of the 120 samples, spectral data from two bands of Near-Infrared (NIR) were taken from each sample at three different stages of ripening. As an alternative, a combination of ANN and the cultural algorithm was used to choose the four most efficient wavelengths. Both methods were successful in achieving high accuracy for low/high acidity classification problems, with 100% for pH and 99.2% for acidity.

1.8.6 Determining the Maturity Indices

The quality and shelf life of fruits in the postharvest value chain is significantly influenced by the stage of fruit maturity at the time of harvest. The application of precise, effective and economical ripeness indices is necessary to provide consumers and processors with high-quality fruits. Fruit ripeness cannot be primarily determined by a single maturity index. A combination of multiple indices is preferable to a single test because it minimizes variability linked to seasonality, location and fruit-to-fruit variation by providing information on the fruit's mature state.

The estimation of the maturity stage in orchards helps to improve the shelf life of fruits. Picking fruits based on their stage of maturity can reduce the cost of storage and increase the market preference.

1.8.7 Monitoring Fruit Freshness

Freshness is the most important fruit quality indicator since it directly affects consumers' preference and purchase intentions especially for climacteric fruits. Additionally, it is a crucial component of market pricing and shelf life. Therefore, it is crucial to research the process for determining fruit freshness. Jiangong et al. [37] suggested that using GoogleNet to track the evolution of banana freshness. They discovered a connection between freshness and shelf life by utilizing transfer learning to analyze the process of freshness changing. Using the GoogleNet model, features of banana photos were automatically extracted and the classifier module subsequently categorized the features. The findings demonstrate that the model is capable of determining whether a banana is fresh and its accuracy level is 98.92%, which is higher than that of a human.

1.8.8 Harvest Ability and Yield

The challenges that our crops will face in the future are unquestionably complex and demanding. Enhancing crop productivity and nutritional security in a sustainable manner will be one of the biggest challenges facing humanity in the next several decades (Scharff et al., [38]).

Fruit traits such as cluster compactness, fruit maturity and berry number per cluster are important to blueberry breeders and producers for making informed decisions about genotype selection related to yield traits and harvestability as well as for plant management. Xueping et al [38] developed a data processing pipeline to count berries, to measure maturity and to evaluate compactness (cluster tightness) automatically using a deep learning image segmentation method for four southern high bush blueberry cultivars ('Emerald', 'Farthing', 'Meadowlark', and 'Star'). An iterative annotation strategy was developed to label images that reduced the annotation time.

1.9 BENEFITS AND CHALLENGES OF AI IN AGRICULTURE

Benefits of AI in agriculture

1.9.1 AI Enables Better Decision-Making

The agricultural sector greatly benefits from predictive analytics. It assists farmers in resolving the major problems associated with farming, like assessing market demands, projecting prices, and determining the best times to plant and harvest crops.

1.9.2 AI Brings Cost Savings

AI-enabled equipment for precision farming enables farmers to grow more crops with less money and resources. AI gives farmers access to real-time information that helps them make informed decisions at every level of farming. This wise choice results in fewer product and chemical loss as well as cost and time savings. Additionally, it enables farmers to pinpoint the precise regions that require pesticide treatment, fertilization, and irrigation, thereby reducing the amount of chemicals used excessively on the crop. In summary, all of these factors lead to lower herbicide usage, higher crop quality, and greater profit with fewer resources.

1.9.3 AI Reduces Labour Shortage

The agriculture sector has long struggled with a workforce deficit. AI can address this problem of farming automation. Examples of how farmers can increase productivity without hiring additional staff include driverless tractors, intelligent spraying, intelligent irrigation and fertilizing systems, vertical farming software, and AI-powered harvesting robots. When compared to human farmhands, AI-driven machinery and equipment are far faster and more accurate.

1.9.4 Challenges of AI Adoption in Agriculture

Given the benefits of AI for sustainable farming, adopting this technology could seem like the right course of action for any farmer. But as everyone knows, there are still some significant obstacles to overcome. They are as follows:

(i) Lack of Familiarity with AI Machines

Even though using AI in agriculture has several advantages, most people worldwide are not aware with the use of AI-enabled tools and solutions. In order to address the problems, AI businesses should first give farmers simple equipment, and then, as they become accustomed to it, more sophisticated machinery.

(ii) Lack of Experience with Emerging Technologies

For underdeveloped nations, implementing AI and other cutting-edge technologies in agriculture might be difficult. In places where such agricultural technology is not being used, it will be extremely difficult to sell such technologies. Farmers in these places need assistance from someone in order to apply these technologies.

(iii) Privacy and Security Issues

AI may give rise to a number of legal concerns because there are currently no defined rules and guidelines for its use. Furthermore, there can be security and privacy problems like cyber attacks and data leaks because of the usage of software and the internet. For farmers or farm owners, all of these problems could pose serious challenges.

1.10 FUTURE THRUST

Future AI technologies will provide precise and creative solutions to the main agricultural problems that face farmers all over the world. Artificial intelligence (AI) will solve nearly every problem, including pest management, weather forecasting and farm labour support. In the upcoming years, there will be more fascinating advancements in AI that have a direct bearing on farming. With the use of data to tailor yields to specific plant rows, artificial intelligence (AI) will help farmers to achieve their target very precisely. AI firms are creating robots that can easily perform a range of tasks in agricultural environments. This robot is designed to harvest crops faster and more thoroughly than a person. These machines are programmed to monitor for weeds and harvest and pack crops while maintaining crop quality.

These robots are also capable of overcoming the challenges that agricultural labour faces.

In order to determine whether an insect has landed and what kind of bug it is—such as a grasshopper or a locust—AI systems use satellite images and historical data. By notifying farmers via text message on their phones to take the necessary actions and apply the proper insect control, artificial intelligence assists farmers in managing pests. In addition to helping farmers automate their operations, AI is advancing precision farming, which uses fewer resources to produce higher-quality and more productive crops. Future technological advancements will support businesses looking to improve AI-based products or services, such as automated machine production, drones and training data for agriculture, helping the world handle food supply issues for an expanding population.

Since most of the most advanced technology is now only used on large, networked farms, the application of AI in agriculture will need to place a strong emphasis on ensuring that everyone can access it. Expanding access and connection to small farms in remote places of the world will assure the future of data science in farming and ML-automated agricultural items. AI will be beneficial and successful in the agriculture sector since it maximizes resource utilization and efficiency and, in large part, addresses the labour and resource deficit. Additionally, this technology will be crucial to horticultural research and development.

1.11 CONCLUSION

The uptake of AI solutions will determine how farming uses AI in the future. Despite ongoing large-scale research and certain commercially available applications, the agricultural industry remains neglected. Furthermore, research is still in its early stages to develop predictive solutions to address genuine problems that farmers experience.

In the future, different types of validated AI algorithms and models with mobile based AI application, test verified with images of commercially important agricultural and horticultural crops, will pave the way for smart agriculture. Adoption of these AI based strategies in farming not only smoothens the different intercultural operations in crops cultivation but also ascertains the efficiency of the works in different crop critical growth stages. Additionally, AI based tools are the need of the hour in the context of "climate change and chemical less farming".

REFERENCES

1. Liu S.Y. Artificial intelligence (AI) in agriculture, *IEEE Transactions on Industrial Informatics*. Vol. 22(3): 14–15, 2020.

2. Talaviya T., Shah D., Patel N., Yagnik H., Shah M. Implementation of artificial intelligence in agriculture for optimization of irrigation and application of pesticides and herbicides, *Artificial Intelligence in Agriculture*. Vol. 4, 58–73, 2020.
3. Redmon J., Divvala S., Girshick R., and Farhadi A., You only look once: unified, real-time object detection, Proc. 2016 IEEE Conf. Computer Vision and Pattern Recognition (CVPR), Las Vegas, NV, USA, Jun. 2016, pp. 779–788, doi: 10.1109/CVPR.2016.91.
4. Tian, Y., Yang, G., Wang, W., Li E., and Liang Z. Detection of Apple Lesions in Orchards Based on Deep Learning Methods of CycleGAN and YOLOV3-Dense. *Journal of Sensors*. Vol. 7630926, 1–7630926,:13, 2019.
5. Liu J., Wang X. Tomato diseases and pests detection based on improved Yolo V3 convolutional neural network. *Frontiers Plant Sci*. Vol. 11: 8982020, 2020. doi: 10.3389/fpls.2020.00898.
6. Wu D., Shuaichao L., Mei J., Huaibo S., Using channel pruning-based YOLO v4 deep learning algorithm for the real-time and accurate detection of apple flowers in natural environments. *Computers and Electronics in Agriculture*. Vol. 178, 105742. 2020. ISSN 0168-1699, https://doi.org/10.1016/j.compag.2020.105742
7. Magalhães S.A., Castro L., Moreira G., dos Santos F.N., Cunha M., Dias J., Moreira A.P. Evaluating the single-shot MultiBox Detector and YOLO deep Learning models for the detection of tomatoes in a hreenhouse, *Sensors* Vol. 21(10): 3569, 2021. https://doi.org/10.3390/s21103569
8. Bazame, H.C., Molin, José P.A., Daniel, M.M. Detection, classification, and mapping of coffee fruits during harvest with computer vision, *Computers and Electronics in Agriculture*. Vol. 183, 2021. 106066, ISSN 0168-1699, https://doi.org/10.1016/j.compag.2021.106066
9. An, K., Wang, Li, Z., Song, C., Tang X. Song, J. Real-time monitoring method of strawberry fruit growth state based on YOLO improved model, in *IEEE Access* Vol. 10, pp. 124363–124372, 2022, doi:10.1109/ACCESS.2022.3220234.
10. Bonora, A., Bortolotti, Gianmarco, B., Kushtrim, C., Luca, G., Luigi M., Reprint of: A convolutianonal neural network approach to detecting fruit physiological disorders and maturity in 'Abbé Fétel' pears *Biosystems Engineering*, Vol. 223, Part B pp. 224–232, ISSN 1537-5110, https://doi.org/10.1016/j.biosystemseng. 2022.10.005.
11. Gai, R., Chen, N. & Yuan, H., A detection algorithm for cherry fruits based on the improved YOLO-v4 model, *Neural Computing and Applications*. Vol. 35, 13895–13906. 2021.
12. Li K., Jiancong W., Hassan J., Hui W. A fast and lightweight detection algorithm for passion fruit pests based on improved YOLOv5. *Computers and Electronics in Agriculture*. Vol. 204,107534.2023. https://doi.org/10.1016/j.compag.2022.107534.
13. Gui Z., Jianneng C., Yang L., Zhiwei C., Chuanyu W., Chunwang D., A lightweight tea bud detection model based on Yolov5, *Computers and Electronics in Agriculture*. Vol. 205, 107636.2023. https://doi.org/10.1016/j.compag.2023.107636.
14. Liu L., Jing L., Jianqing W., Peiyu H., Ling W., Quan Z., An improved YOLOv5-based approach to soybean phenotype information perception, *Computers and Electrical Engineering*. Vol. 106, 108582, 2023. ISSN 0045-7906, https://doi.org/10.1016/j.compeleceng.2023.108582.
15. Paul A., Machavaram R., Ambuj, D.K., Nagar H. Smart solutions for capsicum Harvesting: Unleashing the power of YOLO for Detection, Segmentation, growth stage Classification, Counting, and real-time mobile identification, *Computers and Electronics in Agriculture*. Vol. 219, 2024 ISSN 0168-1699, 108832, ISSN 0168-1699, https://doi.org/10.1016/j.compag.2024.108832.

16. Shi J., Bai Y., Zhou J., Zhang B. Multi-crop navigation line extraction based on improved YOLO-v8 and Threshold-DBSCAN under complex agricultural environments, *Agriculture*. Vol. 14(1): 45, 2024. https://doi.org/10.3390/agriculture14010045
17. Chetan M. Badgujar, Poulose Alwin, Gan Hao. Agricultural object detection with You Look Only Once (YOLO) agorithm: *A Bibliometric and Systematic Literature Review*. Vol. 223, 2024. Article: 109090.
18. Wang X., Zhang X., Cao Y., Wang W., Shen C., Huang T. *SegGPT: Segmenting Everything in Context*. 2023. arXiv:2304.03284. https://doi.org/10.48550/arXiv.2304.03284
19. Kirillov A., Mintun E., Ravi N., Mao H., Rolland C., Gustafson L., Xiao T., Whitehead S., Berg A.C., Lo W-Y., Dollár P., Girshick R. *Segment Anything*. 2023. arXiv:2304.02643. https://doi.org/10.48550/arXiv.2304.02643
20. Kirillov A., Mintun E., Ravi N., Mao H., Rolland C., Gustafson L., Xiao T., Whitehead S., Berg A.C., Lo W-Y., Dollar P., Girshick R. Segment Anything. *2023 IEEE/CVF International Conference on Computer Vision (ICCV)* 2023. 3992–4003. arXiv:2304.02643. https://doi.org/10.48550/arXiv.2304.02643
21. Carraro A., Sozzi M., Marinello F. The Segment Anything Model (SAM) for accelerating the smart farming revolution. *Smart Agricultural Technology*. Vol. 6, 100367, 2023.
22. Gomez D., Selvaraj M.D., Casas J., Mathiyazhagan K., Rodriguez M., Assefa T., Mlaki A., Nyakunga G., Kato F., Mukankusi C., Girma E., Mosquera G., Arredondo V., Espitia E. Advancing common bean (Phaseolus vulgaris L.) disease detection with YOLO driven deep learning to enhance agricultural AI. *Scientific Reports*. Vol. 14, 15596. 2024. https://doi.org/10.1038/s41598-024-66281-w
23. Mandal D., Kumar V., Rao Y.S., An assessment of temporal RADARSAT-2 SAR data for crop classification using KPCA based support vector machine. *Geocarto International*, 1–28. 2020, DOI: https://doi.org/10.1080/10106049.2020.1783577
24. Xue, G., Liu, S., Ma, Y. A hybrid deep learning-based fruit classification using attention model and convolution autoencoder. *Complex Intell. Syst*. Vol. 9, 2209–2219. 2023. https://doi.org/10.1007/s40747-020-00192-x. Published date -12 Oct 2020, Issue Date – June 2023.
25. Asim, M., Ullah, S., Razzaq, A., Qadri, S. Varietal discrimination of guava (Psidium Guajava) leaves using multi features analysis. *International Journal of Food Properties* Vol. 26(1): 179–96. 2022. doi:10.1080/10942912.2022.2158863.
26. Mostafa A.M., Kumar S.A., Meraj T., Rauf H.T., Alnuaim A.A., Alkhayyal M.A. Guava disease detection using deep convolutional neural networks: A case Study of guava plants. *Applied Sciences* 179–96. 2022. https://doi.org/10.3390/app12010239.
27. Cevallos, H., Romero, H., Barrezueta U.S. Application of automatic learning algorithms to classify the fertility of a banana soil. *Universidad y Sociedad*, Vol. 14(5): 29–33, 2020.
28. Elbasi E., Zaki C., Topcu A.E., Abdelbaki W., Zreikat A.I., Cina E., Shdefat A., Saker L. Crop prediction model using machine learning algorithms. *Applied Sciences* 13(16): 9288. 2023. https:// doi.org/10.3390/app13169288.
29. Tirkey D., Singh, K.K., Tripathi, S. Performance analysis of AI-based solutions for crop disease identification, detection, and classification. *Smart Agricultural Technology* Vol. 5, 100238. 2023. ISSN 2772-3755, https://doi.org/10.1016/j.atech.2023.100238.
30. Ramanjot, M.U., Wadhawan A., Singla J., Jhanjhi N.Z., Ghoniem R.M., Ray S.K., Abdelmaboud A. Plant disease detection and classification: A Systematic literature review. *Sensors* Vol. 23(10): 4769. 2023. https://doi.org/10.3390/s2310476930.

31. Yang F, Li F, Zhang, K., ZhangW., and Li S. Influencing factors analysis in pear disease recognition using deep learning. *Peer-to-Peer Networking and Applications* Vol. 14(3): 1816–1828. 2021.
32. Almadhor A., Rauf H.T., Lali M.I.U., Damaševičius R., Alouffi B., Alharbi A. AI-driven framework for recognition of guava plant diseases through machine learning from DSLR camera sensor based high resolution imagery. *Sensors* Vol. 21(11): 3830. 2021. https://doi.org/10.3390/s21113830
33. Sun H., Haowei X., Bin L., Dongjian H., Jinrong H., Haixi Z., Nan G., "MEAN-SSD: A novel real-time detector for apple leaf diseases using improved light-weight convolutional neural networks", *Computers and Electronics in Agriculture*, Vol. 189, 106379. 2021. ISSN 0168-1699, https://doi.org/10.1016/j.compag.2021.106379.
34. Fu, Y., Du, Y., Cao, Z., Li, Q., Xiang, W. A. Deep learning model for network intrusion detection with imbalanced data. *Electronics*, Vol. 11, 898, 2022. https://doi.org/10.3390/electronics11060898.
35. Kader, A.A. Flavor quality of fruits and vegetables, *Science of Food and Agriculture*, 10 June 2008, https://doi.org/10.1002/jsfa.3293.
36. Pourdarbani, R., Sabzi, S., Rohban, M., García-Mateos, G., Paliwal, J., Molina M., José Miguel. Using metaheuristic algorithms to improve the estimation of acidity in Fuji apples using NIR spectroscopy. *Ain Shams Engineering Journal* Vol. 13, 101776. 2022.
37. Jiangong N., Jiyue G., Limiao D., Zhnogzhi H. Monitoring the change process of banana freshness by GoogLeNet, in IEEE Access, Vol. 8 228369–228376, 2020, doi: 10.1109/ACCESS.2020.3045394.
38. Scharff, L.B., Saltenis, V.L.R., Jensen, P.E., Baekelandt, A., Burgess, A.J.,Burow, M., Ceriotti, A., Cohan, J.P., Geu-Flores, F., Halkier, B.A., et al. Prospects to improve the nutritional quality of crops. *Food and Energy Security*. Vol. 11: e327. 2022.
39. Xueping N., Changying L., Huanyu J., Fumiomi T., Ni et al. Deep learning image segmentation and extraction of blueberry fruit traits associated with harvestability and yield. *Horticulture Research* Vol. 7: 110. 2020. https://doi.org/10.1038/s41438-020-0323-3

2 Artificial Intelligence Algorithms and Models for NextGen Agriculture

Dr. Rupali Vairagade, Abhishek Gawali, and Dhanashree Diwase

2.1 INTRODUCTION

The demand for food grows in an era of enormous population growth, challenging the long-running agricultural paradigms which have supported humanity for generations. The strain on traditional farming practices grows as the world's population soars, exceeding the 8 billion mark and projected to attain 9.7 billion by 2050, according to United Nations predictions. There is an instantaneous need for revolutionary solutions as fields that once provided food for a smaller, more controllable population are now unable to meet the growing demands.

Agriculture, which has its roots in age-old customs which have been passed down through the centuries, has historically been the inspiration of human civilization. Equipped with inherited expertise and honed through time, farmers labored the land with a deep attention to its composition, seasonal variations, and agricultural cycles. But just as the foundations of human lifestyles were modified, so did the complexities of survival. Conventional farming strategies, which have been formerly enough for local settlements, are now confronted with the challenge of feeding the growing population[5].

It is important to consider the potential applications of AI in agriculture as we start this investigation. This technological revolution gives farmers the capability to make information-driven selections, going past easy mechanization. AI has the capacity to enhance agricultural control, maximize aid utilization, and decrease dangers related to pests and unpredictable weather conditions. The goal is to enhance production and foster sustainability in the center of agricultural undertakings through the implementation of NextGen farming strategies.

However, this investigation gives a roadmap for answers. This chapter aims to provide light on AI's capability to expand agriculture by addressing the problems of a developing population and the constraints of traditional farming. We will negotiate the interaction of technology and lifestyle through real-world examples, case studies, and a constructive outlook, presenting insights into a future wherein the union of AI and agriculture holds the promise of a resilient, efficient, and sustainable food ecosystem.

DOI: 10.1201/9781003487005-2

2.2 NEXTGEN AGRICULTURE

Next-generation agriculture, or NextGen Agriculture, represents an essential paradigm exchange in the agricultural industry by using cutting-edge technologies to transform conventional farming strategies, especially inside the area of precision agriculture[6]. This innovative approach includes a radical integration of modern-day technologies like robotic systems, artificial intelligence (AI), data analytics, and precision agricultural strategies[5]. One of the key components of NextGen Agriculture is precision agriculture, which stresses the prudent use of technology and data to correctly manipulate field-level activities, enhancing overall efficiency and useful resource allocation.

Precision farming strategies, which strategically employ machine learning algorithms and global positioning systems (GPS) to guide field-level management, are deeply entwined with the NextGen Agriculture technique. With excellent care, farmers can divide their fields into smaller, more manageable zones, each with specific desires primarily based on environmental concerns, crop health, and soil quality. The integration of NextGen Agriculture with precision agriculture is especially clear in terms of the evaluation of past data and cutting-edge situations to forecast best planting, watering, and harvesting seasons[1]. This focused method supports cutting-edge agriculture's sustainable and resource-efficient objectives by means of optimizing yields while lowering the use of water, fertilizers, and pesticides.

The precision agriculture component of NextGen Agriculture is around the widespread applications of data-centric approaches[12]. Using several sensors, drones, and Internet of Things (IoT) gadgets makes it less complicated to gather information in real-time, giving farmers more detailed understanding of crop health, soil properties, and environmental effects. This data-driven strategy significantly aids in useful resource conservation while also enabling farmers to make well-informed decisions on fertilization, irrigation, and insect control. Moreover, the incorporation of machine learning algorithms forms the foundation of predictive modeling, allowing farmers to estimate best planting, irrigation, and harvesting durations with unmatched accuracy.

Drones are vital to NextGen Agriculture and precision agriculture as they offer an airborne angle that is important for quick data collection and analysis[13]. Drones which might be equipped with high-definition cameras and sensors are extremely useful for tracking crop health, diagnosing illness, and evaluating general field conditions. Drone-based surveillance systems can operate more effectively when AI algorithms are included. This is due to the fact robust prediction models may be created and real-time data-driven decision-making is made easier[13].

As the foundation of both NextGen Agriculture and precision agriculture, artificial intelligence orchestrates the decision-making process with an unprecedented level of sophistication. Large datasets feed machine learning algorithms to identify patterns and trends that guide forecasts and suggestions. This functionality makes it feasible to predict crop yields, spot possible disease outbreaks, and allocate resources optimally—all within the precision-driven context of present-day agriculture. Precision agriculture's trademark of minimizing reliance on chemical inputs is aligned with sustainable agricultural practices through the dynamic nature of AI-powered models, which facilitate risk evaluation related to pest infestations[11].

Smart irrigation systems, an important component of precision agriculture, use AI algorithms and sensors to represent the accuracy-cantered method of NextGen Agriculture and precision agriculture. These structures maintain a close eye on weather patterns and soil moisture levels[14], allowing data-driven decision-making to determine the exact amount of water crops need. By strategically avoiding over-irrigation, water resources are preserved, and the power consumption linked to excessive water distribution is reduced. This double gain highlights how NextGen Agriculture and precision agriculture are both resource- and environmentally green.

2.3 MODERN VS TRADITIONAL FARMING

Next-generation agriculture, another name for modern agriculture, is a current farming system that maximizes crop yield through the combination of modern technologies and modern farming strategies. On the contrary, traditional farming describes age-old practices that have been passed down through the generations and are characterized by physical labor and customary ways. While conventional farming has supplied communities for generations, modern agriculture adopts automation, data analytics, and precision agricultural methods.

The contrasts among traditional farming and modern agriculture are extensive and profound. Modern agriculture utilizes cutting-edge technologies like drones, sensors, and artificial intelligence, whereas traditional farming is based on human labor and crude instruments. These traits make it possible to precisely monitor crop health, soil conditions, and climate patterns. This makes it easier to make data-driven decisions to enhance productivity and restrict resource utilization. Aiming to reduce the environmental impact whilst optimizing yield ability, modern agriculture additionally places a strong emphasis on sustainability through strategies like drip irrigation, natural farming, and integrated pest management[4].

The way plants are cultivated, managed, and harvested has been completely transformed with the aid of NextGen Agriculture, which has absolutely revolutionized the agricultural industry[5]. The fusion of modern technology and creative approaches has ushered in a new length of productivity, sustainability, and efficiency. Labor-intensive approaches have been optimized by means of automation and robotics, lowering reliance on physical labor and boosting operational performance. Additionally, the usage of pesticides and fertilizers can be targeted with precision agriculture techniques, decreasing waste and contamination of the environment.

The capability of NextGen Agriculture to conquer the limitations of traditional farming is one among its most crucial contributions. Modern agriculture offers solutions that minimize the problems of environmental degradation and inefficiencies which might frequently be related to traditional farming structures. Using precision farming techniques, farmers may minimize chemical runoff, reduce water utilization, and maximize resource utilization. Furthermore, cutting-edge monitoring systems make it easier to become aware of pests, ailments, and unfavorable weather early on, allowing prompt crop loss prevention measures[3].

The issues of food safety and sustainability are typically met by means of NextGen Agriculture in the face of an increasing global population and growing food demands

[15]. Through growing resilience to weather change, reducing environmental impact, and optimizing productiveness, modern agriculture contributes to securing a strong and sustainable food supply for coming generations. Next-generation agriculture has the capability to significantly impact farming in the future by promoting an environmentally aware, resilient, and efficient agricultural industry through sustained innovation and the implementation of high-quality practices.

2.4 ENHANCEMENT OF TRADITIONAL FARMING USING AI

Traditional farming methods have seen a considerable transition in recent years, primarily due to the integration of Artificial Intelligence (AI) technologies. These cutting-edge trends have converted the agriculture enterprise, reducing reliance on human labor while increasing productivity and sustainability. Traditional agricultural practices have seamlessly transitioned to modern agriculture, also known as NextGen Agriculture, with the aid of making use of AI-driven automation and robotics.

AI-powered automation and robotics have replaced several labor-intensive jobs previously undertaken by farmers, resulting in unrivaled levels of performance and precision. From seeding and planting to irrigation and harvesting, these technologies have greatly reduced physical workload while maximizing resource utilization. For instance, autonomous tractors equipped with AI algorithms can explore fields with precision, optimizing planting patterns and lowering soil compaction.

Moreover, AI-powered tracking devices offer real-time evaluation of crop health, soil conditions, and weather patterns, permitting farmers to make data-driven decisions for better crop control. AI can discover early indicators of pests, ailments, or nutrient deficits using drones and sensors, enabling timely interventions to save crop loss. This amount of understanding and control was previously unbelievable in the context of conventional farming operations.

AI technologies efficiently handle traditional farming's constraints, including resource inefficiency, vulnerability to environmental effects, and labor shortages. By optimizing irrigation schedules, AI facilitates water conservation and reduces waste, addressing one of traditional farming's most essential concerns. Furthermore, AI-powered predictive models can forecast market trends and optimize crop selections and planting procedures, resulting in higher yields and more monetary returns for farmers.

AI addresses some of the restrictions of traditional farming, which include limited access to expert knowledge and information. Farmers in distant or impoverished places frequently lack access to agricultural experts who can recommend high-quality practices and problem-solving techniques. AI-powered systems and mobile applications supply farmers with numerous statistics and information, inclusive of advice for crop control, pest control, and soil health primarily based on actual-time facts and neighborhood situations[3].

The incorporation of AI into conventional farming approaches promises a paradigm shift toward a more efficient, productive, and environmentally friendly agricultural area. Traditional farmers can triumph over long-standing limits by means of leveraging the potential of AI-driven automation, robots, and data analytics to help satisfy the enormous food demands of an ever-increasing populace. As AI advances, its role in enhancing traditional farming methods becomes increasingly important, paving the way for a more robust and profitable agricultural future.

2.5 AI FRAMEWORKS FOR FARMING

Artificial intelligence (AI) systems are increasingly becoming important tools in modern-day agriculture, transforming conventional farming strategies by introducing a new age of statistics-driven decision-making and optimization. Designed particularly for agricultural functions, these systems serve as the spine of innovation, supplying farmers with unparalleled right of entry to modern-day technologies that enhance productiveness, efficiency, and sustainability throughout the rural region.

Central to AI systems are several methodologies and strategies which have been cautiously crafted to address the diverse demanding situations inherent in agriculture. By leveraging the abilities of machine learning, deep learning, reinforcement learning, and decision support systems, those structures enable farmers to navigate the elaborate panorama of current agriculture with great precision and readability.

2.5.1 Machine Learning

Machine learning knowledge of algorithms are instrumental in reading considerable quantities of agricultural statistics to extract valuable insights and styles. In farming, ML techniques are applied to duties inclusive of crop yield prediction, disease detection, soil quality assessment, and pest management [10]. Through supervised and unsupervised learning procedures, ML models are using historic information to make predictions and guidelines that optimize crop management practices and improve universal productivity.

For example, ML algorithms can examine historic weather information, soil composition, and crop characteristics to predict most fulfilling planting instances, decide appropriate irrigation schedules, and identify regions at risk of pest infestations. These insights permit farmers to make knowledgeable choices and optimize resource allocation, in the long run improving crop yields and profitability. ML techniques can also be deployed in precision agriculture packages, in which they examine sensor data from drones and IoT devices to screen crop health, discover anomalies, and discover areas requiring intervention.

2.5.2 Deep Learning

Deep learning techniques, especially convolutional neural networks (CNNs), have transformed agricultural analytics by means of processing massive-scale, high-dimensional data along with satellite imagery, sensor information, and spectral imaging. These models excel in responsibilities like crop classification, weed detection, disease diagnosis, and yield estimation, enabling precise and efficient analysis of agricultural imagery for decision-making purposes.

For instance, DL algorithms can examine satellite views to hit upon early signs of crop strain, discover weed infestations, and estimate crop yields with high accuracy. By leveraging DL models, farmers can proactively control crop health, optimize pesticide use, and forecast yields, leading to advanced useful resource performance and enhanced crop resilience. DL-based total processes additionally facilitate the development of self-sustaining farming structures, wherein robotic gadgets prepared with imaginative and prescient sensors and DL algorithms carry out tasks together with planting, weeding, and harvesting with minimal human intervention.

2.5.3 Reinforcement Learning

Reinforcement learning algorithms give promising possibilities for self-sufficient decision-making in agricultural operations. These algorithms allow agricultural systems, together with robotic harvesters and auto tractors, to examine the most fulfilling techniques with the mode of interacting with their surroundings. Through the feedback of rewards or consequences, reinforcement learning dealers continuously refine their moves, leading to advanced efficiency and useful resource usage in farming practices.

For example, reinforcement learning algorithms can optimize the navigation route of self-reliant tractors based on real-time sensor statistics, soil conditions, and terrain traits. In a comparable vein, RL-based totally robot harvesters can learn how to pick out ripe fruits, navigate via orchards, and optimize harvesting techniques to maximize yield and minimize damage. By incorporating reinforcement learning strategies into agricultural systems, farmers[7] can automate exertion-extensive tasks, lower operational prices, and grow usual farm productivity. Additionally, RL holds capacity for packages which include adaptive irrigation management, in which RL dealers learn to dynamically adjust irrigation schedules based totally on converting environmental situations and crop water necessities.

2.5.4 Decision Support System

Decision help systems integrate AI technologies with domain-particular expertise to aid farmers in making correct selections. These systems analyze numerous datasets, together with climate forecasts, soil situations, marketplace trends, and agronomic expertise, to offer personalized suggestions for crop planning, irrigation scheduling, pest management, and market forecasting. DSS frameworks enable farmers to mitigate dangers, optimize resource allocation [9], and enhance normal farm profitability by means of leveraging superior analytics and predictive modeling talents. For instance, DSS systems can analyze ancient weather data, soil moisture levels, and crop boom styles to recommend the most useful irrigation schedules tailored to particular crop needs and environmental situations.

Additionally[14], DSS frameworks can combine market records and price forecasts to help farmers make strategic selections regarding crop selection, planting schedules, and advertising strategies. By leveraging DSS gear, farmers can optimize aid allocation, decrease dangers, and maximize profits in a more and more complex and dynamic agricultural landscape.

2.6 APPLICATIONS OF AI IN NEXTGEN AGRICULTURE

AI technology is remodeling agriculture by enhancing performance, precision, and sustainability. Key applications include AI in Soil Management, which helps monitor soil health and optimize irrigation; AI in Crop Management, which enables precision farming techniques to enhance yield and resource utilization; and AI in Disease Management, which uses machine learning to detect and predict pest and disease outbreaks for timely intervention.

2.6.1 AI in Soil Management

The software of Artificial Intelligence in soil control optimizes soil fertility, health, and productiveness while lowering environmental impact, consequently revolutionizing agricultural methods. By using AI algorithms, soil records are analyzed to assess fitness indicators like pH, organic matter content, and nutrient stages. This allows farmers to efficiently diagnose deficiencies and compaction troubles. AI-powered precision soil pattern techniques become aware of local heterogeneity in soil traits, enabling centered checking out and analysis for first-class feasible soil amendment and fertilizer utility plans. AI-powered nutrient management structures lessen waste and runoff at the same time as increasing nutrient utilization efficiency with the aid of forecasting nutrient requirements primarily based on crop increase levels, soil nutrient degrees, and meteorological factors. AI-driven real-time soil moisture monitoring maximizes irrigation scheduling to prevent waterlogging or drought strain, enhancing crop output and water performance. Additionally, through figuring out chance regions and placing conservation measures in an area, AI-driven erosion prediction systems help farmers in decreasing the likelihood of soil erosion. Farmers may additionally enhance soil health, improve carbon sequestration, and guarantee long-time period sustainability in agriculture by means of incorporating AI-powered soil management techniques like cowl vegetation and natural amendments.

Table 2.1 presents a top-level view of AI soil management techniques. The device includes an evaluator that selects the excellent choice based on a set of weighted criteria decided by means of the person. Additionally, control-oriented modeling (MOM) is used to lessen nitrate leaching, making sure more sustainable soil management practices. To determine the shortest course between begin nodes and dreams, MOM employs a tactical search approach called "excellent" together with a strategic seek technique known as "hill mountain climbing." Three steps are covered in the engineering know-how to construct the Soil Risk Characterization Decision Support System (SRCDSS): expertise collection, conceptual layout, and device implementation. Sand, clay, and silt contents of soil are inspected by using a synthetic neural network (ANN) version using characteristics from cutting-edge course-decision soil maps together with hydrographic parameters accumulated from a virtual elevation

TABLE 2.1
AI in Soil Management

Application	Technique	Strength	Limitation
[1]	MOM	Minimizes nitrate leaching, maximizes production.	Slow, limited only to nitrogen.
[2]	Fuzzy Logic: SRC-DSS	Can classify soil according to associated risks.	Needs big data. Only a few cases were studied.
[3]	DSS	Reduces erosion and sedimentary yield.	Requires big data for training.
[4]	ANN	Can predict soil enzyme activity. Accurately predicts and classifies soil structure.	Only measures a few soil enzymes. It considers more classification than improving the performance of the soil.
[5]	ANN	Can predict monthly mean soil temperature	Considers only temperature as a factor for soil performance

model (DEM). A better-order neural network (HONN) embeds faraway sensing device measures and characterizes the dynamics of soil moisture.

2.6.2 AI in Crop Management

Crop control with Artificial Intelligence (AI) improves sustainability, performance, and precision throughout the crop production cycle, revolutionizing agricultural methods. Massive volumes of records, including climate patterns, crop traits, and soil conditions, are analyzed by AI-powered systems to offer farmers beneficial statistics for better selection-making. With the help of system-studying algorithms, farmers may additionally maximize yields at the same time as prescribing aid inputs by means of predicting the first-rate times to plant, scheduling irrigation, and identifying regions which can be vulnerable to pest infestations[10]. With unrivaled precision, deep getting-to-know strategies analyze sensor and satellite TV for PC records to pick out sickness outbreaks, weed infestations, and crop stress. This allows for timely interventions and lowers agricultural losses. Robotic agricultural structures are optimized for the usage of reinforcement studying algorithms, taking into consideration accurate and efficient self-sustaining responsibilities such as planting, weeding, and harvesting. Approaches mixed with agronomic understanding to offer custom designed recommendations on market predictions, pest control, and crop making plans, enable farmers to reduce dangers and grow income. Farmers can attain resilient and sustainable agricultural practices, ensuring security and environmental stewardship for future generations, by utilizing AI for crop control.

In Table 2.2, the crop control strategies are compiled. Crop management begins with planting and continues with crop distribution and garage, harvesting, and growth

TABLE 2.2
AI in Crop Management

Application	Technique	Strength	Limitation
[1]	CALEX	Can formulate scheduling guidelines for crop management activities.	Takes time.
[2]	PROLOG	Removes less used farm tools from the farm.	Location specific.
[3]	ANN	Predicts crop yield.	Only captures weather as a factor for crop yield.
[4]	Robotics-Demeter	Can harvest up to 40 hectares of crop	Expensive, uses a lot of fuel.
[5]	Robotics	Has 80% success rate in harvesting crops	Slow picking speed and accuracy.

monitoring. It may be summed up as the endeavors to improve the productivity and enlargement of agricultural merchandise. Crop yield will surely push upward with a thorough comprehension of the elegance of vegetation on the subject of their timing and the kind of soil they thrive in. In order to maximize profitability and protect the surroundings, precision crop control, or PCM, is an agricultural control machine that targets crop and soil inputs according to discipline necessities. Lack of timely, extensively disseminated facts on crop and soil conditions has hindered PCM.

2.6.3 AI in Disease Management

Artificial intelligence (AI) may be used in agriculture to manage plant illnesses through presenting innovative methods to come across, diagnose, and lessen their results, improving crop fitness and yield[2]. Artificial intelligence (AI)-enabled systems look at a lot of statistical resources, which include ancient sickness information, environmental variables, and crop pictures, to exactly pick out ailment styles and signs and symptoms. Early prognosis and focused treatments are made possible through machine studying algorithms that identify and diagnose plant sicknesses based totally on visible symptoms which can be captured using imaging strategies[3]. Large-scale crop fields can benefit from the rapid reaction and containment efforts that deep mastering fashions offer by detecting minor alerts of ailment outbreaks across massive amounts of agricultural pictures, which include drone and satellite statistics.

Reinforcement mastering algorithms facilitate adaptive choice-making and continual development through the years with the aid of optimizing ailment control techniques through gaining knowledge from prior successes and mistakes. AI generation and expert understanding are combined with selection support systems to provide farmers with sensible advice on disorder prevention, control, and remedy

TABLE 2.3
AI in Disease Management

Application	Technique	Strength	Limitation
[1]	Computer vision system (CVS), genetic algorithm (GA), ANN	Works at a high speed. Can multi-task.	Dimension-based detection which may affect good species.
[2]	Rule-Based Expert, Data Base (DB)	Accurate results in the tested environment.	Inefficacy of DB when implementing on a large scale.
[3]	Fuzzy Logic (FL), Web GIS	Cost effective, eco-friendly.	Inefficiency due to scattered distribution. Takes time to locate and disperse data. The location of the data is determined by a mobile browser.
[4]	FL Web-Based, Web-Based Intelligent Disease Diagnosis System (WIDDS)	Good accuracy. Responds swiftly to the nature of crop diseases.	Limited usage as it requires internet service. Its potency cannot be ascertained as only 4 seed crops were considered.

(Table 2.3). This increases crop resilience and lowers economic losses. Farmers can take pre-emptive steps to guard their vegetation, assure food safety, and enhance sustainable agricultural practices by utilizing AI to manage sicknesses.

A farmer ought to use an inclusive sickness management and control technique that consists of bodily, chemical, and organic measures with a view to effectively control ailments and limit losses[4]. It takes a lot of time and money to do these, which is why the usage of an AI approach to sickness manipulation and management is necessary.

2.7 FUTURE SCOPE OF AI IN AGRICULTURE

AI's ability applications in agriculture are significant and feature the ability to completely remodel farming strategies and solve fundamental issues that the world is now going through. AI-powered advanced precision agriculture will develop similarly, combining drones, satellite TV for PC images, and Internet of Things sensors to display vegetation and environmental situations with formerly unheard-of accuracy. With this advancement, farmers might be capable of making statistics-driven choices in actual-time that optimize the usage of assets, cut waste, and increase yields in big-scale operations. Drones for crop spraying, robotic harvesters, driverless tractors, and AI-powered self-reliant farming systems will all be crucial to simplifying agricultural operations and boosting output, efficiency, and protection. Artificial intelligence (AI)-powered predictive analytics will permit farmers to foresee and decrease weather-related dangers. Platforms for virtual agriculture will become the focal factors for records integration, evaluation, and choice aid. These hubs will give farmers admission to insightful advice and recommendations for better farm management. AI will also remodel genetic improvement and crop breeding, accelerating the new development of recent varieties with desirable traits and supporting the advancement of climate-smart agriculture. Furthermore, via giving smallholder farmers an entry to the scalable and fairly priced era, AI has the capability to empower them and boost their livelihoods, resilience, and manufacturing. All matters taken into consideration, AI's destiny in agriculture is shiny, because it has the power to result in innovative shifts toward more resilient, equitable, and sustainable agricultural systems that guarantee both the sustainability of the surroundings and food security for coming generations.

2.8 CONCLUSION

Apart from the formerly cited advantages, the integration of AI in agriculture has the capability to sell more diversity and accessibility among farmers. Artificial intelligence (AI) has the ability to democratize agricultural innovation and enable smallholder farmers to improve their livelihoods and resilience towards market volatility and climate change by giving them access to AI-powered tools and technology, irrespective of their region or assets.

Furthermore, it is expected that as AI in farming maintains to strengthen, it will spur interdisciplinary partnerships and collaborations, encouraging a lifestyle of innovation and cross-area information sharing. To fully utilize AI in addressing complex and demanding agricultural situations and starting up new avenues for sustainable meals production, numerous professional and attitude integrations are essential, from partnerships between farmers and technology companies to agronomists and statistics scientists.

Moreover, ethical considerations and suitable deployment techniques are getting increasingly important as AI generation becomes superior and widely available. Building trust and self-belief among stakeholders and keeping the long-term viability and integrity of agricultural structures depend upon assuring the moral use of AI in agriculture, which addresses troubles like records privacy and safety, algorithmic bias, and transparency[14].

REFERENCES

1. Li, M. and Yost, R.S., 2000. Management-oriented modeling: optimizing nitrogen management with artificial intelligence. Agricultural Systems, 65(1), pp.1–27.
2. López, E.M., García, M., Schuhmacher, M. and Domingo, J.L., 2008. A fuzzy expert system for soil characterization. Environment International, 34(7), pp.950–958.
3. Montas, H. and Madramootoo, C.A., 1992. A decision support system for soil conservation planning. Computers and Electronics in Agriculture, 7(3), pp.187–202.
4. Tajik, S., Ayoubi, S. and Nourbakhsh, F., 2012. Prediction of soil enzymes activity by digital terrain analysis: comparing artificial neural network and multiple linear regression models. Environmental Engineering Science, 29(8), pp.798–806.

5. Levine, E.R., Kimes, D.S. and Sigillito, V.G., 1996. Classifying soil structure using neural networks. Ecological Modelling, 92(1), pp.101–108.
6. Plant, R.E., 1989. An artificial intelligence based method for scheduling crop management actions. Agricultural Systems, 31(1), pp.127–155.
7. Lal, H., Jones, J.W., Peart, R.M. and Shoup, W.D., 1992. FARMSYS—a whole-farm machinery management decision support system. Agricultural Systems, 38(3), pp.257–273.
8. Pilarski, T., Happold, M., Pangels, H., Ollis, M., Fitzpatrick, K. and Stentz, A., 2002. The demeter system for automated harvesting. Autonomous Robots, 13, pp.9–20.
9. Van Henten, E.J., Hemming, J., Van Tuijl, B.A.J., Kornet, J.G., Meuleman, J., Bontsema, J. and van Os, E., 2002. An autonomous robot for harvesting cucumbers in greenhouses. Autonomous Robots, 13(3), pp.241–258.
10. Eli-Chukwu, N.C., 2019. Applications of artificial intelligence in agriculture: A review. Engineering, Technology & Applied Science Research, 9, pp.1–4.
11. Jesus, J., Panagopoulos, T.. and Neves, A., 2008, June. Fuzzy logic and geographic information systems for pest control in olive culture. In 4th IASME/WSEAS International Conference on Energy, Environment, Ecosystems & Sustainable Development. Retrieved from www.researchgate.net/.
12. Kolhe, S., Kamal, R., Saini, H.S. and Gupta, G.K., 2011. A web-based intelligent disease-diagnosis system using a new fuzzy-logic based approach for drawing the inferences in crops. Computers and Electronics in Agriculture, 76(1), pp.16–27.
13. Kolhe, S., Kamal, R., Saini, H.S. and Gupta, G.K., 2011. An intelligent multimedia interface for fuzzy-logic based inference in crops. Expert Systems with Applications, 38(12), pp.14592–14601.
14. Munirah, M.Y. and Rozlini, M., 2013, October. An expert system development: its application on diagnosing oyster mushroom diseases. In 2013 13th International Conference on Control, Automation and Systems (ICCAS 2013) (pp. 329–332). IEEE.

3 Precision Agriculture
Challenges and Opportunities

Aaskaran Bishnoi, Gurwinder Singh, Ranjit Singh, and Simrat Walia

3.1 INTRODUCTION

Precision agriculture, also known as precision farming or smart farming, represents a revolutionary approach to agricultural management that harnesses the power of technology and data analytics to optimize farming practices [1]. At its core, precision agriculture seeks to enhance efficiency and productivity by precisely tailoring inputs such as water, fertilizers, pesticides, and seeds to meet the unique requirements of individual areas within a field or crop. By applying inputs in the right amounts, at the right time, and in the right locations, precision agriculture endeavors to maximize yields while minimizing resource wastage and environmental harm.

This approach marks a significant departure from traditional farming methods, which often involve uniform application of inputs across entire fields regardless of variations in soil composition, moisture levels, or crop health. By contrast, precision agriculture relies on a combination of advanced technologies, including satellite imagery, drones, sensors, GPS technology, and farm machinery equipped with sensors, to collect data on various factors influencing crop growth and health. This data is then analyzed using sophisticated algorithms and software to generate actionable insights that guide decision-making in farming operations. One of the key principles of precision agriculture is site-specific management, which involves treating different areas of a field individually based on their unique characteristics. For example, soil types, moisture levels, nutrient levels, and topography can vary significantly within a single field. By leveraging data-driven insights, farmers can tailor their management practices to address these variations more effectively. This may involve adjusting irrigation schedules, applying fertilizers or pesticides only where and when needed, and adopting precision planting and harvesting techniques to optimize yield potential.

The adoption of precision agriculture holds tremendous promise for revolutionizing modern farming practices and addressing the challenges of feeding a growing global population while minimizing environmental impacts. By optimizing resource use and minimizing waste, precision agriculture can help farmers achieve higher yields and profitability while reducing their ecological footprint.

DOI: 10.1201/9781003487005-3

3.1.1 Potential Benefits

Here's an overview of how precision agriculture works and its potential benefits:

1. Data Collection and Analysis: Precision agriculture relies heavily on data collection from various sources such as satellite imagery, drones, sensors, GPS technology, and farm machinery equipped with sensors. This data is then analyzed using advanced algorithms and software to gain insights into soil health, crop growth, pest infestations, weather patterns, and other relevant factors [2].
2. Site-Specific Management: One of the key principles of precision agriculture is site-specific management. Instead of treating an entire field uniformly, farmers can use data-driven insights to tailor their management practices to specific areas within the field. For example, areas with different soil types, moisture levels, or nutrient levels may require different amounts of irrigation, fertilizers, or pesticides [3].
3. Optimized Resource Use: By precisely targeting inputs based on real-time data and analysis, farmers can optimize the use of resources such as water, fertilizers, and pesticides. This not only reduces input costs but also minimizes waste and environmental pollution.
4. Improved Crop Yields: Precision agriculture enables farmers to better monitor crop health and respond quickly to any issues that arise. By identifying problems early, such as pest infestations, nutrient deficiencies, or diseases, farmers can take timely corrective actions, leading to healthier crops and higher yields. Additionally, site-specific management allows for more precise planting, seeding, and harvesting techniques, further enhancing yield potential [4].
5. Environmental Sustainability: Perhaps one of the most significant benefits of precision agriculture is its potential to enhance environmental sustainability. By reducing the use of agrochemicals and minimizing waste, precision agriculture helps mitigate negative impacts on soil, water, and air quality.

3.1.2 Evolution of Precision Agriculture

Precision agriculture has evolved significantly over time, with its roots dating back to ancient agricultural practices [5]. Precision agriculture has undergone a remarkable evolution throughout history (Table 3.1), transitioning from traditional farming practices to the advanced techniques [6,7]. Early agricultural methods relied heavily on manual labor and observational knowledge, with ancient civilizations like the Sumerians and Egyptians pioneering early irrigation systems to enhance crop yields [8]. The advent of mechanical farming equipment in the 20th century, such as tractors and plows, marked a significant milestone, revolutionizing agriculture and improving productivity. This era saw the adoption of variable rate application techniques and the integration of GPS-guided equipment to enhance precision in farming practices. Farmers began leveraging advanced analytics and predictive modeling techniques to optimize crop management and resource allocation, leading to more efficient farming

TABLE 3.1
Evolution of Precision Agriculture

Era	Technology	Characteristics
1970s–1980s	Early GPS systems	Basic GPS technology used for mapping fields and navigation.
1990s	GIS (Geographic Information Systems)	Integration of spatial data for mapping soil variability, yield monitoring, and decision support systems.
2000s	Remote Sensing	Utilization of satellite and aerial imagery for crop health monitoring, disease detection, and yield prediction.
2010s	IoT (Internet of Things)	Integration of sensors, drones, and telemetry for real-time monitoring of soil moisture, temperature, and more.
2020s	AI and Machine Learning	Implementation of AI algorithms for predictive analytics, precision spraying, autonomous machinery, and more.

practices [9,10]. Looking ahead, the future of precision agriculture is poised to incorporate emerging technologies like IoT, artificial intelligence, and blockchain, paving the way for even more precise and sustainable farming methods [11,12].

3.2 TECHNOLOGICAL LANDSCAPE

Precision agriculture, also known as precision farming or smart farming, leverages technology to optimize the efficiency, productivity, and sustainability of agricultural practices [13] (see Figure 3.1). The technological landscape for precision agriculture is diverse and rapidly evolving. Here's an overview.

1. Remote Sensing and Imaging: Satellite imagery, drones (Unmanned Aerial Vehicles UAVs), and ground-based sensors provide detailed data on crop health, soil moisture levels, and other vital parameters. This data is used for monitoring, mapping, and decision-making.
2. GIS (Geographic Information Systems): GIS technology integrates various data sources, such as soil maps, weather forecasts, and crop yield data, into a spatial framework. It helps farmers analyze and interpret data to make informed decisions about field management.
3. Data Analytics and AI (Artificial Intelligence): Advanced analytics techniques, including machine learning and AI algorithms, process large datasets to extract valuable insights. AI-powered systems can predict crop yields, detect diseases, optimize irrigation, and recommend personalized agronomic practices.
4. Precision Irrigation Systems: Automated irrigation systems, equipped with sensors and controllers, deliver water precisely where and when it's needed.

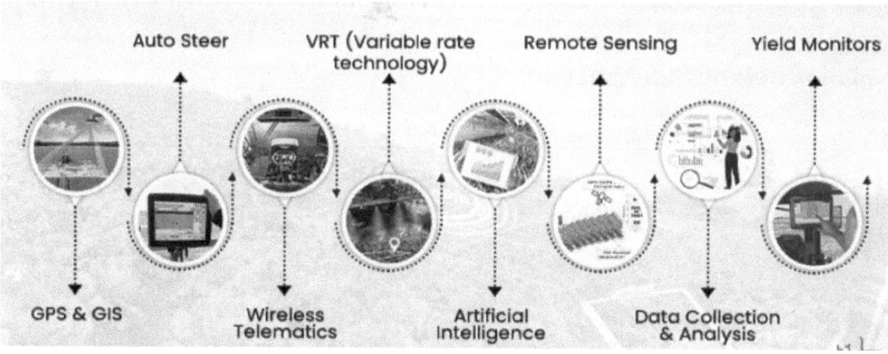

FIGURE 3.1 Block Diagram for Components of Precision Farming.

TABLE 3.2
Summary Statistics

	healthy	multiple_diseases	rust	scab
count	1821.000	1821.000	1821.000	1821.000
mean	0.283	0.050	0.342	0.325
std	0.451	0.218	0.474	0.469
min	0.000	0.000	0.000	0.000
25%	0.000	0.000	0.000	0.000
50%	0.000	0.000	0.000	0.000
75%	1.000	0.000	1.000	1.000
max	1.000	1.000	1.000	1.000

This minimizes water wastage, reduces irrigation costs, and enhances crop yields.

5. Variable Rate Technology (VRT): VRT enables farmers to apply inputs such as fertilizers, pesticides, and seeds at variable rates across a field, based on spatial variability in soil properties and crop requirements. This optimizes resource usage and improves crop uniformity.

3.3 MACHINE LEARNING MODELS

The dataset consists of plant pathology data [14], where each sample is classified into one of two classes: Healthy and Disease. In total, there are 1821 samples in the dataset. The majority of samples belong to the Disease class, with 1305 samples, while the Healthy class contains 516 samples. To facilitate model training and evaluation, the dataset is split into a training set and a test set. The training set comprises 1456 samples, while the test set contains 365 samples.

Table 3.2 displays summary statistics for four variables: “healthy”, “multiple diseases”, “rust”, and “scab”. Each row represents a different statistic, including count, mean, standard deviation, minimum, 25th percentile, median (50th percentile), 75th percentile, and maximum. The values in the table provide information such as the number of observations (count), the average value (mean), the spread of the data (standard deviation), the range of values (minimum and maximum), and the quartiles. This summary table offers a concise overview of the distribution of each variable in the dataset, aiding in understanding their central tendency and variability.

Table 3.3 represents a correlation matrix showing the correlation coefficients between different variables. Each row and column in the table represents a variable, namely "healthy", "multiple diseases", "rust", and "scab". The values within the table indicate the correlation between pairs of variables. For example, the correlation coefficient between "healthy" and "rust" is approximately -0.453, suggesting a moderately negative correlation between these two variables. Similarly, the correlation coefficient between "multiple diseases" and "rust" is approximately -0.165, indicating a weaker negative correlation.

Table 3.4 represents the performance metrics of a Random Forest Classifier model for a binary classification task. For the class labeled as "0", the precision is 1.00, indicating that when the model predicts the class as "0", it is correct 100% of the time. The recall is 0.48, suggesting that the model correctly identifies 48% of all actual instances of class "0". The F1-score, which is the harmonic mean of precision and recall, is 0.65. The support column indicates that there are 245 instances of class "0" in the dataset. Similarly, for the class labeled as "1", the precision is 0.49, indicating that when the model predicts the class as "1", it is correct 49% of the time. The recall is 1.00, suggesting that the model correctly identifies all instances of class "1". The F1-score is also 0.65. The support column indicates that there are 120 instances of

TABLE 3.3
Correlation Matrix

	healthy	multiple_diseases	rust	scab
healthy	1.000	-0.144	-0.453	-0.436
multiple_diseases	-0.144	1.000	-0.165	-0.159
rust	-0.453	-0.165	1.000	-0.500
scab	-0.436	-0.159	-0.500	1.000

TABLE 3.4
Classification Report

	Precision	Recall	F1-score	Support
0	1.00	0.48	0.65	245
1	0.49	1.00	0.65	120
Accuracy			0.65	365
Macro avg	0.74	0.74	0.65	365
Weighted avg	0.83	0.65	0.65	365

TABLE 3.5
SVM Classification Report

	Precision	Recall	F1-score	Support
0	0.73	1.00	0.85	265
1	1.00	0.04	0.08	100
Accuracy			0.74	365
Macro avg	0.87	0.52	0.46	365
Weighted avg	0.81	0.74	0.64	365

class "1" in the dataset. The overall accuracy of the model is 0.65, indicating that it correctly predicts 65% of the instances in the dataset.

Table 3.5 presents the classification report for an SVM model used for a binary classification task. For class "0", the precision is 0.73, indicating that when the model predicts instances as class "0", it is correct approximately 73% of the time. The recall is 1.00, suggesting that the model correctly identifies all instances of class "0". The F1-score, which balances precision and recall, is 0.85. The support column indicates there are 265 instances of class "0" in the dataset. For class "1", the precision is 1.00, indicating that when the model predicts instances as class "1", it is correct 100% of the time. However, the recall is 0.04, indicating that the model only correctly identifies 4% of all actual instances of class "1". Consequently, the F1-score for class "1" is only 0.08. The support column shows there are 100 instances of class "1" in the dataset. The accuracy of the model is 0.74, meaning it correctly predicts the class label for 74% of instances in the dataset. The macro average of precision, recall, and F1-score is 0.87, while the weighted average is 0.81, indicating the model's performance when considering the support for each class. Overall, the table provides insights into the SVM model's performance metrics for both classes, along with overall accuracy and average scores.

Figure 3.2 is a graph of a receiver operating characteristic (ROC) curve. A ROC curve is a visual representation of the performance of a binary classification model at various thresholds. It helps assess how well the model can distinguish between positive and negative cases. The X-axis of the ROC curve represents the False Positive Rate (FPR), which is the proportion of negative cases incorrectly classified as positive. The Y-axis represents the True Positive Rate (TPR), which is the proportion of positive cases correctly classified as positive. The diagonal line in the ROC curve represents a random classifier. A random classifier would have an AUC of 0.5, which means it performs no better than random guessing. The curves in the image represent the performance of three different classification models: Random Forest, Logistic Regression, and Random Classifier. The area under the curve (AUC) is a metric that summarizes the performance of a classification model. The AUC of a model can range from 0 to 1. A higher AUC indicates better performance. In the image, the Random Forest model has the highest AUC (0.74), followed by Logistic Regression (0.74) and Random Classifier (0.5). This means that the Random Forest model performs the best at distinguishing between positive and negative cases, followed by Logistic Regression and then the Random Classifier (Figure 3.3).

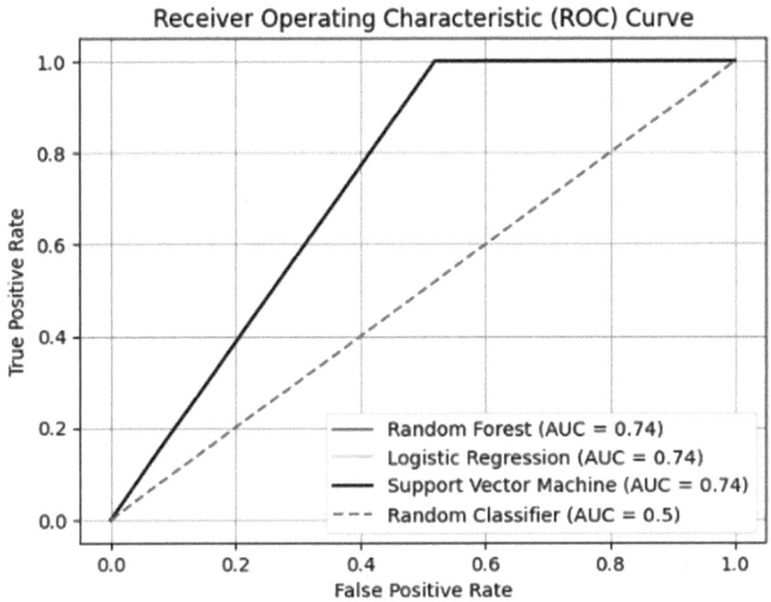

FIGURE 3.2 ROC Curve Comparison for Three Models.

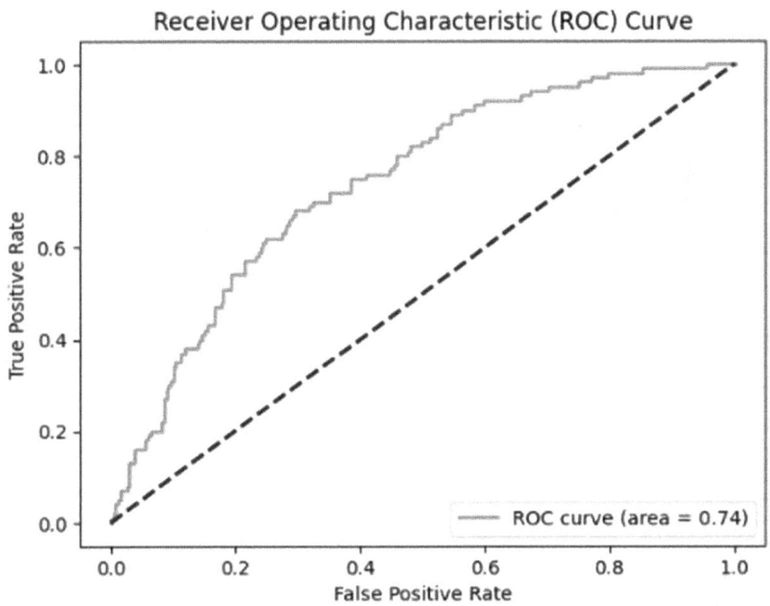

FIGURE 3.3 ROC Curve Comparison.

3.4 CHALLENGES AND BARRIERS

There are major challenges to precision agriculture, they include; high costs for technology purchases and maintenance, lack of access to installation, high complexities in handling the data, and lack of skilled personnel among farmers. These challenges contribute to the lack of widespread adoption and, therefore, hinder effective implementation, particularly by low-income stakeholders.

1. Cost of Implementation: One of the primary barriers to the widespread adoption of precision agriculture is the high initial cost of implementing advanced technologies and systems [15]. Investments in hardware, software, sensors, and machinery can be substantial, particularly for small-scale farmers or those operating on tight budgets.
2. Access to Technology and Infrastructure: Limited access to technology and infrastructure poses a significant challenge for many farmers, particularly those in remote or underserved regions [16]. Inadequate internet connectivity, unreliable power supply, and lack of technical support can hinder the adoption and effective utilization of precision agriculture tools and services.
3. Knowledge and Skills Gap: Successfully implementing precision agriculture practices requires farmers to have a solid understanding of agronomy, data science, and technology [17]. However, many farmers lack the necessary knowledge and skills to effectively utilize advanced agricultural technologies and interpret data-driven insights.
4. Compatibility and Interoperability Issues: The interoperability of different precision agriculture technologies and systems is often limited [18]. Integrating disparate systems and ensuring seamless communication between hardware and software components can be challenging, especially when dealing with proprietary solutions from multiple vendors.
5. Risk and Uncertainty: Transitioning to precision agriculture involves inherent risks and uncertainties [15]. Factors such as weather variability, market fluctuations, and technical failures can impact the success and ROI of precision agriculture investments.

3.5 OPPORTUNITIES AND BENEFITS

- Increased Efficiency: Precision farming optimizes resource usage by applying inputs such as water, fertilizers, and pesticides precisely where and when they're needed, reducing waste and improving efficiency (Figure 3.4).
- Enhanced Productivity: By leveraging data-driven insights and advanced technologies, precision farming techniques can lead to higher crop yields and improved farm productivity.
- Environmental Sustainability: By minimizing chemical usage, reducing soil erosion, and conserving water resources, precision farming practices promote environmental sustainability and help mitigate the impact of agriculture on ecosystems.

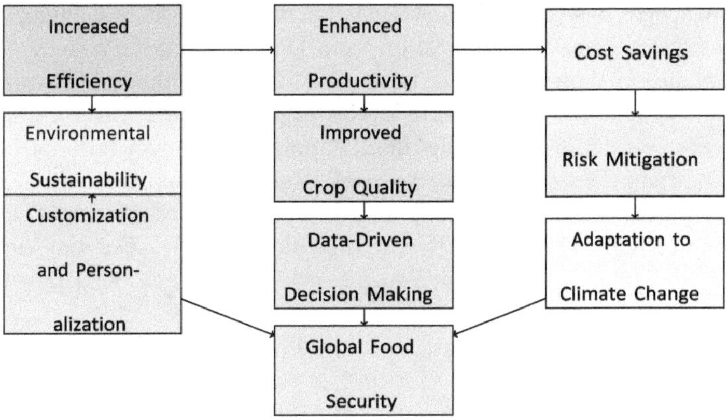

FIGURE 3.4 Block Diagram for Opportunities and Benefits of Precision Farming.

- Improved Crop Quality: Precision farming enables farmers to monitor and manage crop health more effectively, resulting in better quality produce with fewer defects and higher market value.
- Customization and Personalization: Precision farming technologies allow for customized agronomic practices tailored to the specific needs of each field or even individual plants, maximizing productivity and sustainability.
- Adaptation to Climate Change: Precision farming techniques help farmers adapt to changing climate conditions by providing tools and strategies to mitigate the effects of extreme weather events, droughts, and other climate-related challenges.

3.6 CASE STUDIES

FarmERP – Integrated Farm Management Software
FarmERP is an integrated farm management software developed by Shivrai Technologies, based in Pune, India [19]. It offers comprehensive solutions for managing various aspects of agricultural operations, including crop planning, inventory management, financial analysis, and traceability. FarmERP has been adopted by several large-scale farming enterprises and agribusinesses across India, including Jain Irrigation Systems Ltd. and Sahyadri Farms. These organizations have integrated FarmERP into their operations to streamline processes, improve productivity, and enhance decision-making. By leveraging FarmERP, farmers and agribusinesses have achieved significant improvements in operational efficiency, resource utilization, and yield optimization. For example, Sahyadri Farms, one of India's largest grape producers, reported a 20% increase in yield and a 15% reduction in input costs after implementing FarmERP.

Tata Kisan Sansar – Digital Agriculture Platform
Tata Kisan Sansar is a digital agriculture platform launched by Tata Chemicals, aimed at empowering smallholder farmers in rural India [20]. The platform provides

farmers with access to weather forecasts, market prices, agronomic advice, and financial services through mobile applications and localized service centers. Tata Kisan Sansar has been rolled out in several states across India, including Maharashtra, Uttar Pradesh, and Karnataka. The platform leverages partnerships with local agricultural extension services, cooperatives, and input suppliers to deliver tailored support and services to farmers. The Tata Kisan Sansar platform has enabled smallholder farmers to make informed decisions about crop planning, input procurement, and market sales, leading to improved productivity and profitability. By accessing timely information and advisory services, farmers have reduced risks associated with weather variability and market uncertainties.

Weather-Based Crop Insurance Scheme – Pradhan Mantri Fasal Bima Yojana (PMFBY)

The Pradhan Mantri Fasal Bima Yojana (PMFBY) is a weather-based crop insurance scheme launched by the Government of India to protect farmers against crop losses due to natural calamities and adverse weather conditions. The scheme utilizes remote sensing and satellite technology to assess crop damages and facilitate timely insurance payouts. PMFBY has been implemented across all states and union territories of India, covering millions of farmers and hectares of agricultural land. The scheme is administered by the Ministry of Agriculture and Farmers Welfare [21] in collaboration with state governments, insurance companies, and financial institutions. PMFBY has provided financial security to farmers and incentivized adoption of risk-mitigation practices such as crop diversification and improved agricultural practices. By compensating farmers for crop losses, the scheme has helped stabilize incomes and reduce vulnerabilities associated with climate-related risks.

3.7 SUSTAINABLE AND PRODUCTIVE FUTURE

The sustainable and productive future of precision farming envisions a transformative shift in agricultural practices towards greater efficiency, resilience, and environmental stewardship. In this future scenario, precision farming plays a central role in addressing key challenges facing the agricultural sector while unlocking new opportunities for sustainable growth and development. Here are some key aspects that characterize the sustainable and productive future of precision farming:

1. Efficiency and Productivity: Precision agriculture optimizes resource usage, leading to increased efficiency and productivity in farming practices. By applying inputs precisely where and when they're needed, waste is minimized, and crop yields are enhanced.
2. Cost Savings and Economic Viability: Through the adoption of precision agriculture technologies, farmers can reduce input costs and maximize returns on investment. This not only improves economic viability for farmers but also contributes to the overall sustainability of agricultural operations.
3. Environmental Sustainability: Precision agriculture promotes environmental sustainability by minimizing chemical usage, reducing soil erosion, conserving

water resources, and mitigating the impact of agriculture on ecosystems. By adopting more efficient and sustainable farming practices, precision agriculture helps preserve natural resources for future generations.

4. Quality and Safety of Agricultural Products: By enabling farmers to monitor and manage crop health effectively, precision agriculture ensures the production of high-quality agricultural products with fewer defects and higher market value. This enhances food safety and consumer confidence in agricultural products.
5. Risk Mitigation and Resilience: Advanced monitoring and predictive analytics in precision agriculture help farmers identify and mitigate risks such as pest infestations, diseases, and adverse weather conditions. This improves the resilience of agricultural systems to external shocks and uncertainties.
6. Innovation and Technological Advancements: Precision agriculture continues to evolve with advancements in technology, including artificial intelligence, robotics, and data analytics. These innovations hold the promise of further revolutionizing the agricultural sector and addressing emerging challenges such as climate change and food security.
7. Collaboration and Knowledge Sharing: The success of precision agriculture relies on collaboration among farmers, researchers, industry stakeholders, and policymakers. By sharing data, insights, and best practices, stakeholders can collectively innovate and drive positive change in the agricultural industry.

3.8 CONCLUSION

Precision farming enhances farm efficiency, productivity and the sustainable use of inputs through technological applications and information management. Growing from the age-old conventional practices to the technological era involving computing, GPS, and machine learning, the future of precision agriculture looks even more promising. Comparing two models using machine learning, namely the Random Forest Classifier and Support Vector Machine in plant pathology unveiled that SVM offered the best accuracy of 0.74 and the Random Forest model has a test accuracy of 0.65. While there are some shortcomings such as high costs, limited accessibility, issues with data processing and storage, and lack of knowledge, precision agriculture provides more efficiency, productivity, and economic and environmental revolution, and better crop quality. The success stories include John Deere's Precision Planting System and India's PMFBY are the examples for this. Future developments are expected to increase productivity, cost effectiveness, better sustainability practices, and exchange of knowledge among participants in all agriculture activities with high levels of accuracy and efficiency.

REFERENCES

1. Gebbers, R., and V. I. Adamchuk. 2010. Precision agriculture and food security. *Science* 327 (5967): 828–831.
2. He, P., and J. Jin. 2016. Precision agriculture—a worldwide overview. *Computers and Electronics in Agriculture* 124: 3–6.

3. Srinivasan, A. 2014. A review on crop yield prediction methods. *International Journal of Agricultural and Biological Engineering* 7 (2): 51–56.
4. Mueller, T. G. 2018. Precision agriculture: A challenge for crop nutrition management. In *Nutrient Dynamics for Sustainable Crop Production*, eds. J. Benbi, T. Daniel, and M. S. Aulakh, 295–319. Springer.
5. Smil, V. 2011. Harvesting the biosphere: The human impact. *Population and Development Review* 37 (4): 613–636.
6. Bailey, D. 2018. Agriculture in ancient Egypt. *Ancient History Encyclopedia*. Retrieved from www.ancient.eu/article/1216/agriculture-in-ancient-egypt/
7. Gebbers, R., and V. I. Adamchuk. 2010. Precision agriculture and food security. *Science* 327 (5967): 828–831.
8. He, P., and J. Jin. 2016. Precision agriculture—a worldwide overview. *Computers and Electronics in Agriculture* 124: 3–6.
9. Srinivasan, A. 2014. A review on crop yield prediction methods. *International Journal of Agricultural and Biological Engineering* 7 (2): 51–56.
10. Blackmore, S., and R. J. Godwin. 2003. Precision agriculture: A pragmatic approach. In *Proceedings of the 5th European Conference on Precision Agriculture*, 17–24.
11. Mueller, T. G. 2018. Precision agriculture: A challenge for crop nutrition management. In *Nutrient Dynamics for Sustainable Crop Production*, ed. J. Benbi, T. Daniel, and M. S. Aulakh, 295–319. Springer.
12. Shen, J., J. Liu, Q. Zhou, P. Shi, and Z. Jiang. 2020. Precision agriculture technologies for sustainable and efficient crop production. *Sustainability* 12 (14): 5702.
13. Kitchen, N. R., K. A. Sudduth, S. T. Drummond, P. C. Scharf, H. L. Palm, and D. F. Roberts. 2010. Precision agriculture technologies for crop management. *Journal of Soil and Water Conservation* 65 (4): 111A–117A.
14. Kaggle. n.d. Plant Pathology 2020: Logistic Regression vs KMeans. Retrieved from www.kaggle.com/code/amrkhaledsaleh/plant-pathology-2020-logistic-regression-vs-kmeans.
15. O'Connell, S., and M. Brennan. 2020. Challenges and opportunities for the adoption of precision agriculture technologies: A systematic literature review. *Precision Agriculture* 21 (2): 390–416.
16. Lowenberg-DeBoer, J. 2001. Barriers to the adoption of precision farming systems. *Journal of Crop Production* 3 (2): 371–396.
17. Arshad, M., S. Aziz, and Z. Nisar. 2019. Barriers to adoption of precision agriculture: A study of Pakistani farmers. *Journal of Agricultural Science and Technology* 21 (4): 937–950.
18. Wolfert, S., L. Ge, C. Verdouw, and M. J. Bogaardt. 2017. Big data in smart farming – A review. *Agricultural Systems* 153: 69–80.
19. Shivrai Technologies. 2022. FarmERP – Integrated Farm Management Software. Retrieved from www.farmerp.com/
20. Tata Chemicals. 2022. Tata Kisan Sansar – Digital Agriculture Platform. Retrieved from www.tatakisansansar.com/
21. Ministry of Agriculture and Farmers Welfare, Government of India. 2022. Pradhan Mantri Fasal Bima Yojana (PMFBY). Retrieved from https://pmfby.gov.in/

4 Plant Disease Detection Using Machine Learning

Garima Singh, Ravi Shankar Singh, Shweta Kushwaha, Prasanth Kumar Bevara, Romir Mehta, and Vicky Kumar Nayak

4.1 INTRODUCTION

The farming industry is essential for worldwide aliment and the production of food. Still, the health and productivity of crops are constantly threatened by various diseases affecting plants. Earlier farmers were using common methods of disease detection and classification which are often tedious and may not be convincing in preventing the rapid spread of diseases. In this era, there has been a rise in research focusing on integrating techniques in machine learning for the timely and precise identification of plant diseases. Here, we dig into the evolution made in detecting and classifying plant diseases using machine learning methods. The combination of artificial intelligence, particularly with machine learning, has demonstrated potential in transforming plant disease management. For this context, we present a systematic review by [1], which discusses innovative methods for identifying and categorizing plant diseases through the application of machine learning. The article, "Machine Learning Using SVM and Other Kernel Techniques" [2], delves into various machine learning methods, highlighting the importance of support vector machines (SVM) and other kernel techniques. Comparative analysis of deep learning and traditional machine learning techniques in plant leaf disease detection are described in [3]. Exploratory research showcases instances of employing K-means clustering and K-NN-based classification, as delineated in the study by [4], including the introduction of convolutional neural networks (CNN), Bayesian optimized SVMs, and stochastic hybrid method forest group [5]. Also discussed are specific applications in the context of different crops, such as groundnut leaf disease detection using the back propagation algorithm [6], grape leaf disease detection employing SVM classifiers [7], and wheat leaf image classification based on the presence of rust disease using fuzzy C-means clustering [8].

The study aims to identify plant diseases, which are a serious risk to food security and global agriculture. A thorough grasp of these disorders' foundations is necessary for their efficient management. Our work is outlined in Section 4.2 describes the essentials of plant pathology. Section 4.3 discusses the comparison between traditional methods and machine learning methods. Section 4.4 presents the data utilized in this investigation, encompassing the entirety of the data preprocessing techniques employed. Section 4.5 describes the different models used in the study. Section 4.6

DOI: 10.1201/9781003487005-4

presents the evaluation metrics utilized to assess the model, with a performance analysis covered in Section 4.7. Lastly, Section 4.8 addresses the conclusion and potential future directions of this study.

4.2 ESSENTIALS OF PLANT PATHOLOGY

This section discusses the essentials of plant pathology as precision agriculture with emphasis on basic plant diseases, types, symptoms, and signs.

4.2.1 Basics of Plant Diseases

Diseases affecting plants are a major concern for global agriculture, profoundly impacting crop yields and food security. Understanding plant diseases is crucial for effective management. They can be categorized by the nature of the pathogens involved: Fungal diseases, such as rusts and blights, thrive in humid conditions and affect various plant tissues, causing discoloration, wilting, and tissue death. These pathogens spread through airborne spores, direct contact, or contaminated soil and water. Bacterial diseases, including bacterial wilt and leaf spot, spread via contaminated soil, water, or plant debris and enter plants through natural openings or wounds, causing wilting, spotting, and rot. Viral diseases, like mosaic and ring spot viruses, are often transmitted by insect vectors, resulting in mottling, stunting, and malformed growth. Other pathogens include nematodes, which attack plant roots and cause stunted growth and nutrient deficiencies, and phytoplasmas, bacteria-like organisms affecting the vascular system, leading to yellowing and distorted growth. These pathogens spread through various modes, such as direct contact, airborne spores, contaminated soil and water, or vectors like insects. Understanding these transmission modes is essential for effective disease prevention and control strategies, enabling farmers and researchers to develop targeted approaches to mitigate the impact of these illnesses on crop health and productivity.

4.2.1.1 Types of Plant Diseases

Pathogens causing **infectious diseases** pose a significant risk to plant health, manifesting in various ways across different categories. Fungal diseases, including powdery mildew, rusts, and smuts, lead to discoloration, lesions, or a powdery coating on the surfaces of plants [9],[10]. Bacterial pathogens, responsible for ailments like bacterial blight and leaf spot, result in wilting, necrosis, or water-soaked lesions [11]. Viral diseases, including tobacco mosaic virus and cucumber mosaic virus, produce mosaic patterns, leaf distortion, and stunted growth [12]. Plant-parasitic nematodes, like the root-knot nematode, cause diseases by forming galls on plant roots. **Non-infectious diseases,** stemming from environmental factors, include abiotic stress with symptoms like wilting and chlorosis, and pollution-related disorders from air or soil pollutants, causing issues such as stippling on leaves due to ozone pollution [13],[14]. Physiological disorders, like blossom-end rot in tomatoes, result from disruptions in physiological processes, often exacerbated by soil moisture fluctuations [15],[16]. Understanding these diseases' distinct characteristics is crucial for effective plant disease management strategies.

4.2.2 Symptoms and Signs

Understanding the symptoms and signs associated with various plant diseases is crucial for effective model training and accurate identification. The manifestations and indicators may differ significantly based on the crop variety and the particular ailment. Some examples are as follows:

4.2.2.1 Leaf Spotting and Lesions

Leaf spot symptoms appear as brown, white, or black spots with rings and dark edges on leaves, varying by plant, pathogen, and development stage. Fungal bodies, resembling black rocks, form rings within these spots and can merge into cysts. Diagnosis involves observing these symptoms and fungal bodies. [17] used digital image processing to detect palm oil leaf diseases from leaf lesions.

4.2.2.2 Discoloration and Yellowing

Abnormal leaf discoloration, known as chlorosis, indicates a problem in chlorophyll production, resulting in yellowing leaves due to the loss of green pigment. This condition can stem from nutrient deficiencies (like nitrogen, iron, or magnesium), water scarcity, environmental factors (such as extreme heat or insufficient light), diseases, and pests. Chlorosis typically starts at the leaf tips or edges and spreads inward, impairing photosynthesis and inhibiting the plant's ability to produce energy and essential nutrients. Identifying the cause of chlorosis is crucial for taking corrective actions to restore plant health. A study [6] investigated groundnut leaf disease detection, highlighting discoloration as a prominent symptom.

4.2.2.3 Rust Formation

Rust-colored powdery spores on the surface of leaves, a common plant affliction, manifest as reddish or orange dust on leaves, stems, and fruit due to infection by Pucciniales fungi. These fungi invade plant tissues, establishing a parasitic relationship in moist environments ideal for their spread through wind-borne spores. Upon contacting a host, the spores penetrate and grow within plant cells, disrupting normal functions like photosynthesis and nutrient uptake, leading to stunted growth, yellowing leaves (chlorosis), and potentially plant death. Effective rust management involves crop rotation, cleanliness, and fungicide application. A study [7] used fuzzy C-means clustering to classify wheat leaves based on rust presence.

4.2.2.4 Curling and Twisting

Curling or twisting of leaves indicates physiological stress. Curling involves bending or curving inward or outward, often due to temperature, moisture changes, or applied stress. Twisting involves rotational deformation along an axis, resulting in a spiral or coil caused by torsional forces or uneven tension distribution. Both phenomena affect structural integrity, functionality, and aesthetics. [18] used deep learning for image-based detection of plant diseases, recording symptoms like curling.

4.2.2.5 Necrosis and Wilting

Necrosis and wilting are commonly observed phenomena in plants experiencing stress or disease, with necrosis involving the premature death of tissue leading to browning or blackening, often due to environmental stressors, pathogen attacks, or physical damage, and wilting manifesting as drooping or limpness of leaves or stems caused by insufficient water uptake or loss of turgor pressure due to water stress or vascular diseases. Both serve as indicators of plant distress, aiding in diagnosing and addressing underlying health issues. [5] employs a hybrid feature-based approach for detecting plant diseases, focusing on symptoms like necrosis.

4.2.2.6 Blight and Rot

"Sudden wilting, browning, and rotting of plant tissues" are symptoms of plant diseases like sugarcane leaf scorch, characterized by the rapid deterioration of leaves. Researchers used image processing techniques to identify this condition, including K-means clustering for segmentation and K-nearest neighbors (K-NN) for classification. K-means clusters pixels based on color and intensity, isolating regions that may indicate leaf scorch. K-NN then classifies these regions as healthy or diseased. This approach targets symptoms like blight, aiming for accurate detection and management of the disease [3].

4.2.2.7 Deformities and Malformations

Abnormal growth patterns in plants, such as twisted stems or deformed leaves, can significantly impact health and productivity. Causes include genetic mutations, environmental stresses, pests, diseases, nutrient deficiencies, chemical exposure, or extreme weather. Understanding these causes is crucial for plant health and agricultural yield. [18] used convolutional neural networks to identify plants, considering these deformities.

4.2.2.8 Mottling and Mosaics

Distinct discoloration patterns, mottling, or mosaic-like formations on leaves often indicate viral infections, nutritional deficiencies, or other stress factors. These symptoms disrupt the uniform green coloration of healthy leaves, impacting photosynthesis and overall plant health. In plant disease diagnosis, [19] uses machine learning techniques to analyze spectral data and detect these symptoms.

4.2.2.9 Stunted Growth

Stunted plant growth indicates potential issues like disease, nutrient deficiency, and environmental stress. Such plants fail to reach full size, reducing yield and quality. [20] introduced a semi-automated system for detecting leaf diseases in soybean crops, aiding disease control. Evaluating plant health involves analyzing physical traits like color, size, and form. As used by [21], innovative phenotyping methods employ affordable 3D technologies for precise plant measurements, enabling early identification of health problems and targeted interventions for improved growth and yield.

4.2.3 Importance of Early Detection

This section elaborates the importance of early detection in plant diseases and their economic impact followed by the available facilities.

4.2.3.1 Significance of Detecting Plant Disease at an Early Stage

Early detection of plant diseases is essential because symptoms can be subtle and not immediately visible. Catching diseases early allows for prompt intervention, which prevents the problem from worsening. This proactive approach helps control disease spread, ensures healthier crops, and supports sustainable agriculture by reducing the need for more aggressive treatments later on [22].

4.2.3.2 Economic Impacts of Delayed or Undetected Diseases

Unidentified or postponed recognition of plant diseases can profoundly affect the economy. They lead to diminished crop production by affecting the plant's ability to perform photosynthesis, uptake nutrients, and generate robust fruits or grains. Such a decline in productivity results in reduced earnings for farmers and a decrease in the quality of agricultural produce. This subsequently causes a drop in market value and adds to the economic hardship. Prompt identification of plant diseases is essential to lessen these effects, safeguard the yield and quality of crops, and ensure the financial well-being of farmers.

4.2.3.3 Early Detection Facilitates Prompt and Targeted Disease Management

Early detection in plant disease management enables targeted interventions, reducing the need for broad-spectrum treatments that can harm non-target organisms and the environment. By identifying diseases early, farmers can implement precise measures that limit economic losses, minimize agrochemical use, and promote sustainable agricultural practices. This approach not only preserves environmental health but also enhances the resilience and productivity of agricultural systems [22].

4.3 TRADITIONAL VS. MACHINE LEARNING METHODS

This section highlights the advantages of machine learning models over the traditional methods.

4.3.1 Limitations of Traditional Methods

Traditional plant disease detection methods face several challenges, prompting the shift to advanced technologies like machine learning. These conventional approaches often need more subjective visual inspections [1], time-consuming field surveys, and manual processes [3], which are impractical for large areas. They may also lack scalability [4], require specialized expertise [5], and struggle with early detection and quantitative analysis [8],[17]. Additionally, these methods can be affected by environmental factors, pose a risk of disease spread through physical contact [6], and have limited automation, reducing efficiency in large-scale monitoring [12].

4.3.2 Advantages of Machine Learning in Precision Agriculture

Machine learning enhances precision agriculture by revolutionizing farming practices and improving efficiency. It aids early disease detection through analysis of plant images and environmental data [1], boosts precision agriculture with insights into crop health and weather patterns [3], and refines crop yield predictions by examining historical and weather data [5]. Machine learning also supports weed and pest management by distinguishing crops from weeds, reducing pesticide use [8],[17], automating harvesting and sorting to cut labor costs [6], and enhancing soil health monitoring and resource management [12],[23]. It further improves climate resilience through predictive models and provides personalized guidance for crop management [13],[18]. Machine learning helps farmers make data-driven decisions, boosting productivity and efficiency [16],[20].

4.4 DATA COLLECTION AND PREPROCESSING

This section illustrates the various repositories of plant databases that are publicly available for the practitioners in this field and it also summarizes various data preprocessing techniques.

4.4.1 Sources of Plant Databases

4.4.1.1 Public Repositories

Publicly accessible datasets from repositories like Kaggle, the University of California Irvine machine learning repository (UCI-ML repository), or specialized agriculture databases offer a diverse range of datasets. These datasets, comprising images, sensor data, or other relevant information, serve as valuable resources for plant disease detection studies [12],[23].

4.4.1.2 Collaboration with Agricultural Institutions

Collaborating with agricultural research institutions or farms establishes opportunities for accessing real-world data. This collaboration may involve on-field data collection using sensors, cameras, or other data-gathering tools, ensuring a direct link to practical agricultural scenarios.

4.4.2 Data Preprocessing Techniques

4.4.2.1 Data Cleaning

Ensuring dataset quality relies heavily on data preprocessing, with data cleaning being a key element. Proper handling of missing values and outliers is crucial.

Missing values can introduce biases and errors, which can be addressed through imputation or omitting affected records. Outliers can skew results, so identifying and removing them improves dataset reliability. Following data cleaning practices, as outlined by [5] and [11], ensures high-quality data for accurate analysis and insights.

4.4.2.2 Feature Engineering

Feature engineering is crucial for developing machine learning models in plant disease detection. It involves selecting and transforming relevant characteristics from raw data using domain knowledge and statistical methods. Essential techniques include normalization and scaling: normalization adjusts numerical features to a standard range (0 to 1) to prevent any feature from disproportionately affecting the model. In contrast, scaling ensures features with larger magnitudes don't overshadow those with smaller ones, aiding model stability and convergence. In studies like [4] and [18], normalization and scaling were essential for effective deep learning and preprocessing in disease detection models.

4.4.2.3 Data Augmentation (for Image Data)

Methods for augmenting image data, including rotation, flipping, and cropping, enhance the variety within a dataset by generating new variations of existing images. These transformations help generate multiple image versions, improving model generalization. For example, rotating or flipping an image can produce different versions, while cropping focuses on various regions. This process of artificially diversifying the dataset through data augmentation is particularly beneficial for tasks such as image-based plant disease detection [12],[13], where having a large and diverse dataset is crucial for training accurate and robust models.

4.4.2.4 Handling Imbalanced Datasets

Balancing datasets is crucial for effective machine learning in plant disease detection. Imbalance occurs when one class (e.g., healthy plants) dominates, leading to poor performance detecting the minority class (e.g., diseased plants). To address this, resampling methods like oversampling (duplicating minority class instances) and undersampling (reducing majority class instances) are used to balance the dataset. [5] notes that these techniques improve the model's ability to classify healthy and diseased plants accurately.

4.4.2.5 Splitting the Datasets

Splitting the dataset is a crucial step in developing machine learning models. Three subsets are created: training set, validation set, and test set.

Training Set: It is used to train the model. The majority of the dataset (70–80 %) is allocated for training.

Validation Set: Used to fine-tune the model's parameters and compare different model architectures. Helps to avoid overfitting.

Test Set: Completely independent datasets use unseen data to evaluate model performance. Provides an unbiased estimate of the model's generalization ability.

4.4.2.6 Addressing Data Privacy Concerns

Adopting protective measures is crucial given the rising concerns about data privacy, especially with sensitive information. Anonymization and encryption are key strategies. Anonymization removes or alters identifying information to hide individual identities, while encryption encodes data to prevent unauthorized access. Using these techniques helps protect sensitive data, reduce data breach risks, and build trust among stakeholders, enabling safe data use within privacy regulations [20],[24].

4.4.2.7 Documentation Metadata

Maintaining detailed documentation of data provenance, including sources, collection methods, and preprocessing steps, ensures transparency and reproducibility of the research. This documentation is crucial for facilitating a clear understanding of the dataset's history and processing steps [5],[10].

4.4.3 Description of Dataset

Figure 4.1 offers an overview of the diverse datasets utilized in this study for detecting plant diseases, while Table 4.1 presents a summary of the three datasets used. In this study, we employ three distinct datasets to improve plant disease detection: the Plant Village dataset [www.kaggle.com/datasets/abdallahalidev/plantvillage-dataset (08/06/2023), the Rice Leaf Diseases dataset [www.kaggle.com/datasets/dedeikhsandwisaputra/rice-leafs-disease-dataset (08/06/2023), and the Wheat Leaf Diseases dataset [www.kaggle.com/datasets/buffyhridoy/wheat-disease (08/06/2023).

The Plant Village dataset, sourced from Kaggle, encompasses a vast collection of 70,295 images representing 38 classes across 15 plant species. This dataset includes both healthy and diseased crops such as tomato and apple. From the Plant Village dataset, we extract data for two specific crops: potato and tomato. Additionally, the Rice Leaf Diseases dataset, also obtained from Kaggle, consists of 2,628 images categorized into six classes, providing a balanced representation of various rice leaf diseases and healthy conditions. Similarly, the Wheat Leaf Diseases dataset, sourced from Kaggle, comprises 2,160 images categorized into three classes, focusing on diseases like Brown Rust and Yellow Rust in wheat leaves.

Combining these datasets enables the creation of a comprehensive and diverse dataset, encompassing a wide range of plants, diseases, and healthy conditions. This aggregated dataset serves as a robust foundation for training deep-learning models for effective plant disease detection.

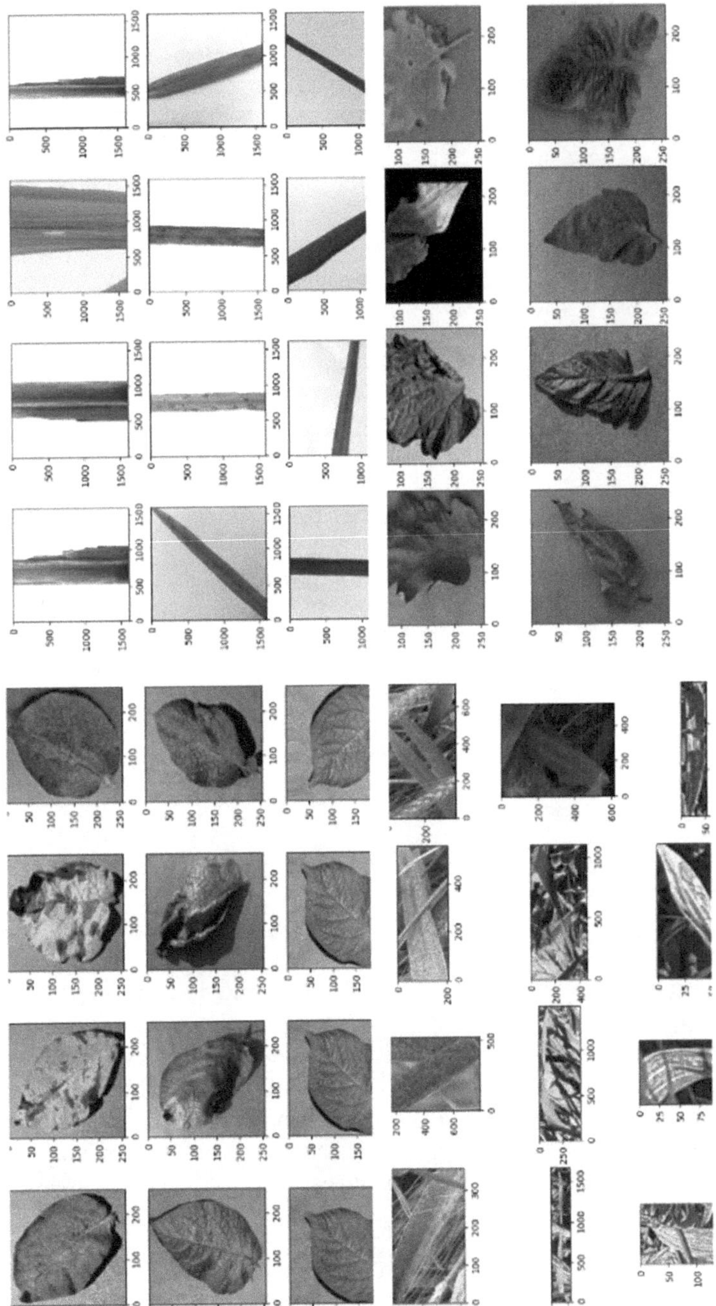

FIGURE 4.1 Used Crop Images.

TABLE 4.1
Description of Dataset

S.No.	Plant Name	Healthy or Diseased	Disease Name	Images (in number)
1	Potato	Diseased	Early_blight	1939
2		Diseased	Late_blight	1939
3		Healthy	Healthy	1824
4	Tomato	Diseased	Bacterial_spot	1702
5		Diseased	Early_blight	1920
6		Diseased	Late_blight	1851
7		Diseased	Leaf_Mold	1882
8		Diseased	Septoria_leaf_spot	1745
9		Diseased	Spider_mites Two-spotted_spider_mite	1741
10		Diseased	Target_Spot	1827
11		Diseased	Tomato_Yellow_Leaf_Curl_Virus	1961
12		Diseased	Tomato_mosaic_virus	1790
13		Healthy	Healthy	1926
14	Rice	Diseased	Bacterial_Leaf	438
15		Diseased	Brown_Spot	438
16		Healthy	Healthy	438
17		Diseased	Leaf_Blast	438
18		Diseased	Leaf_Scald	438
19		Diseased	Narrow_brown_Rust	438
20	Wheat	Diseased	Wheat_Brown_Rust	902
21		Diseased	Wheat_Yellow_Rust	924
22		Healthy	Healthy	334
Total				28835

4.5 MACHINE LEARNING MODELS FOR DISEASE DETECTION

In plant disease detection, machine learning models automate the identification and classification of diseases using data-driven methods. By analyzing patterns in images or sensor data, these models, including SVM, CNN, Decision Trees, Naïve Bayes, and AdaBoost, provide swift and accurate diagnoses. "Machine learning, which focuses on developing models for prediction or decision-making, is generally divided into supervised and unsupervised learning. Each category employs distinct algorithms that provide unique advantages in detecting diseases."

4.5.1 Support Vector Machines (SVM)

SVM is a supervised learning suitable for both classification and regression tasks. When applied to plant disease detection, it adeptly discriminates between healthy and diseased plants by analyzing their characteristics extracted from either images or sensor data, as evidenced by studies [5],[2],[7],[14],[24]–[26].

SVM aims to find a hyperplane represented by Equation 1 [2] that maximally separates the classes while minimizing the classification error. Support Vector Machines are powerful classifiers that work by finding a hyperplane in the feature space that best separates different classes.

$$w * x + b = 0 \tag{1}$$

Given a set of training samples $(\mathbf{x}_1, y_1)$, $(\mathbf{x}_2, y_2)$, ..., $(\mathbf{x}_n, y_n)$ where $\mathbf{x}_i$ is a feature vector and y_i is the corresponding label (disease or healthy).

The optimization problem for SVM is formulated as shown in Equation 2

$$min_{w,b} \frac{1}{2}\|w\|^2 \mid C\sum_{i=1}^{n} \varepsilon_i \tag{2}$$

Subject to constraint shown in Equations 3 and 4

$$(\mathbf{w} \bullet \mathbf{x}_i + b) \geq 1 - \xi_i \tag{3}$$

$$\$xi_i \geq 0 \tag{4}$$

Here, C is a regularization parameter, and ξ_i are slack variables allowing for some misclassification. SVM was chosen for its efficient implementation and performance and has proven to be excellent for high-dimensional problems and small datasets. SVM constructs hyperplanes within an n-dimensional space to train input vectors, enabling effective classification. This hyperplane has the maximum distance to the nearest training data point for all classes (function margin). Thus, the extracted (color features and texture features) features for each image are first normalized and then trained using an SVM classifier whose best-fit parameters are found using GridSearchCV in Python.

4.5.2 Decision Trees

Decision trees are used for classification based on a set of decision rules. In plant disease detection, decision trees can be employed to make decisions about the health status of plants based on input features [11]. Decision trees split the data based on feature values to make decisions. At each node, a decision is made based on a feature, leading to different branches. The process continues until a stopping condition is met. The split at each node is determined by finding the feature and threshold that maximizes information gain (for classification) or variance reduction (for regression).

4.5.3 Random Forest Classifier

Random Forest, an ensemble learning algorithm, enhances classification accuracy by combining predictions from multiple decision trees, which reduces overfitting and improves robustness [5],[27]. This method aggregates tree outputs to classify instances or compute predictions, addressing the overfitting issue common in individual decision trees. Although Random Forest may introduce a slight increase in bias, it generally offers better accuracy with a higher number of trees. It is valued for its robust performance, simplicity, and speed in machine learning applications.

4.5.4 Convolutional Neural Networks (CNN)

In the field of plant disease detection, Convolutional Neural Networks (CNNs) emerge as dynamic tools, demonstrating a capacity to learn intricate features from images autonomously, facilitating precise classification of healthy and diseased plants [3],[5],[6],[9],[12],[13],[18],[23],[25],[28],[29]. The architecture of a CNN includes key components such as convolutional layers, pooling layers, and fully connected layers. In a simplified scenario considering grayscale images denoted as I, a convolution operation, expressed in Equation 5, convolves the image I with a kernel K, generating an output feature map S.

$$S(i, j) = (I * K)(i, j) \tag{5}$$

The ReLU activation function is expressed by the formula in equation 6, which introduces non-linearity. Subsequently, a pooling operation downsamples the feature map, and a fully connected layer outputs the result through a softmax function.

$$A(i, j) = max(0, S(i, j)) \tag{6}$$

Building upon this foundation, we introduce a custom attention mechanism named 'SoftAttention' into the CNN architecture to refine its focus on specific regions of input feature maps, thereby enhancing the model's capability to capture significant details in images. The SoftAttention layer employs 3D convolution to compute attention weights, which are determined through a softmax activation. The layer can either aggregate attention maps or not based on the configuration. Optionally, the attention mechanism concatenates the input with the scaled input according to attention weights. This mechanism empowers the model to selectively attend to different parts of the input data, effectively highlighting areas crucial to the task at hand.

Furthermore, we integrate the Inception ResNet V2 model into this enhanced CNN framework. The Inception ResNet V2, a pre-trained deep neural network architecture, boasts exceptional performance in image classification tasks. In our proposed model, we leverage the Inception ResNet V2 model loaded using TensorFlow's Keras API. Feature maps extracted from a specific layer of this model referred to as conv, are then utilized in tandem with the SoftAttention layer. The SoftAttention mechanism, with its attention-augmented feature maps, is coupled with the powerful

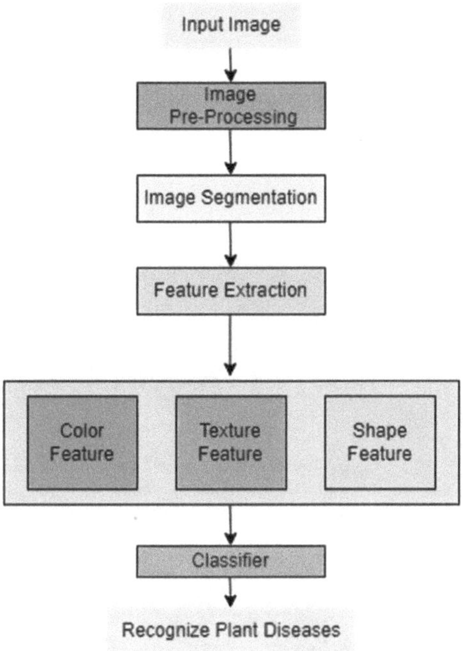

FIGURE 4.2 General Model Architecture.

feature extraction capabilities of the Inception ResNet V2 model. This fusion aims to achieve superior performance in image classification tasks by combining the attention mechanism's fine-grained feature selection and the robust feature extraction abilities of the Inception ResNet V2 model.

The general model architecture is illustrated in Figure 4.2, offering a visual representation of the components and their organization within the model. The diagram likely illustrates various key elements such as input layers, hidden layers, and output layers. The purpose of using this diagram is to provide an overview of the structural arrangement of the model, serving as a reference for understanding the overall design without delving into the specifics of connections or dependencies between individual components. Also, the architecture is illustrated in Figure 4.3, showcasing the inception ResNetV2 with a soft attention block. This innovative combination holds promise for elevating image recognition and classification accuracy in the domain of plant disease detection.

4.6 EVALUATION METRICS FOR MODEL PERFORMANCE

The concept of performance evaluation metrics for models is crucial for assessing how well a model generalizes to new, unseen data. One of the most common methods used for metrics is accuracy; also, depending on the nature of the problem, additional metrics may be more informative. Here are some performance evaluation metrics that can be used for the problems.

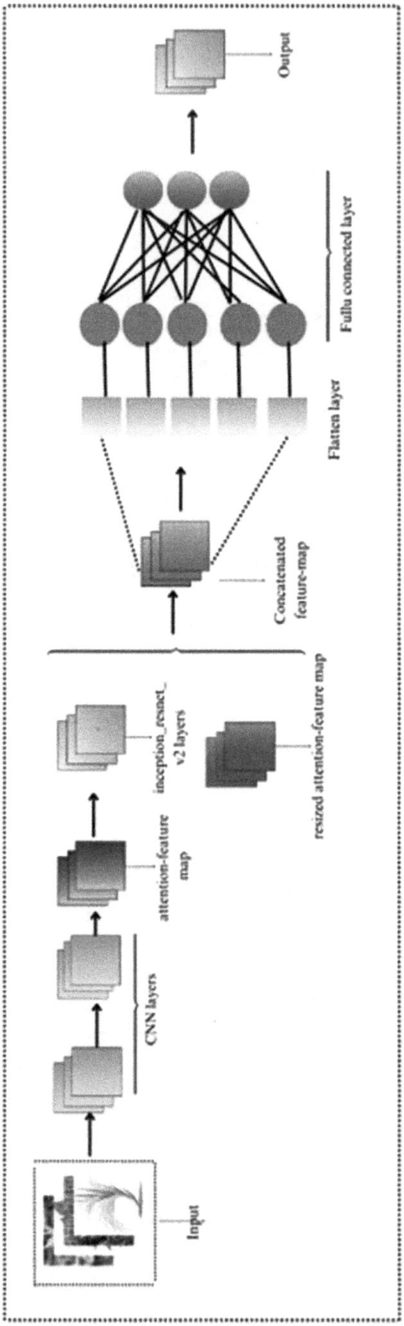

FIGURE 4.3 Architecture of CNN with Soft-Attention Mechanism and Inception Resnet V2.

4.6.1 Accuracy

Equation 7 shows how accurately the model performs by considering true positives and true negatives [30].

$$Accuracy = \frac{Number\,of\,Correct\,Predctions}{Total\,Number\,of\,Predictions} \tag{7}$$

4.6.2 Precision

Equation 8 measures the accuracy of the model's optimistic predictions by indicating the percentage of correctly identified positive instances among all instances predicted as positive [31].

$$Precision = \frac{True\,Positive}{True\,Positive + False\,Positive} \tag{8}$$

4.6.3 Recall

Equation 9 quantifies the Recall metric, which assesses the model's capacity to accurately identify all positive instances. It signifies the correlation of actual positive instances correctly predicted by the model [31].

$$Recall = \frac{True\,positive}{True\,Positive + False\,Negative} \tag{9}$$

4.6.4 F1-Score

Equation 10 denotes the F1 score, a metric calculated as the harmonic mean of precision and recall. This measure considers false positives and false negatives to provide a balanced evaluation [31].

$$F1 - Score = 2\frac{Precision.Recall}{Precision + Recall} \tag{10}$$

4.6.5 Area Under the Receiver Operating Characteristic (ROC) Curve (AUC-ROC)

"The ROC curve depicts the correlation between the true positive rate and the false positive rate, as specified in equation 11. The term AUC-ROC refers to the Area Under the ROC Curve, quantifying the curve's covered area. This metric serves to assess the model's capability in accurately differentiating between positive and negative instances [32]."

$$AUC - ROC = \int_0^1 TPR(FPR^{-1}(t)(dt)) \quad (11)$$

4.6.6 Mean Squared Error (MSE) for Regression Models

Equation 12 in regression models computes the mean squared disparity between predicted (y^i) and observed (yi) values [33].

$$MSE = \frac{1}{n}\sum_{i=1}^{n}(y_i - y_i)^2 \quad (12)$$

4.7 MODEL PERFORMANCE ANALYSIS

The comparison of machine learning models, shown in Table 4.2, indicates that the Decision Tree achieved the highest validation accuracy at 98.66%, followed by the Random Forest at 96.72% and SVM at 94.39%. This highlights the effectiveness of ensemble methods like Decision Trees and Random Forests. Table 4.3 provides a comparison between CNN models utilizing Soft-attention and the Inception ResNet v2 architectures. The Inception ResNet v2 consistently demonstrates superior performance compared to the Soft-attention CNN in all crop categories. Specifically, it achieves an accuracy of 94.10% for tomatoes, while the Soft-attention CNN reaches 91.30%. Figure 4.4 details these findings, with (a) focusing on the performance of machine learning models, (b) on pre-trained models for different crops, and (c) on the accuracy and power loss of CNN architectures.

TABLE 4.2
Comparison of Machine Learning Models

Model	Validation	Accuracy
SVM	94.39	93.37
Decision Tree	98.66	98.48
Random Forest	96.72	99.96

TABLE 4.3
Comparison with Some Pre-Trained Model

Crop	VGG16	Inception V3	AlexNet	ResNet50	CNN
Potato	85.00%	89.50%	86.20%	80.80%	91.30%
Tomato	91.20%	93.40%	90.70%	88.60%	94.10%
Rice	79.50%	82.10%	80.40%	76.90%	83.70%
Wheat	87.80%	90.20%	88.70%	85.30%	91.80%

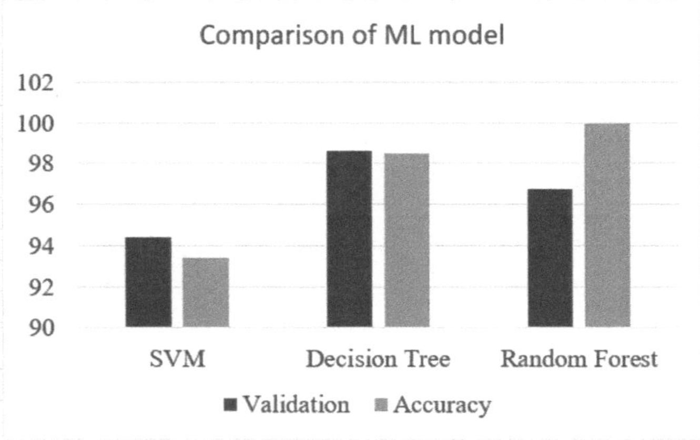

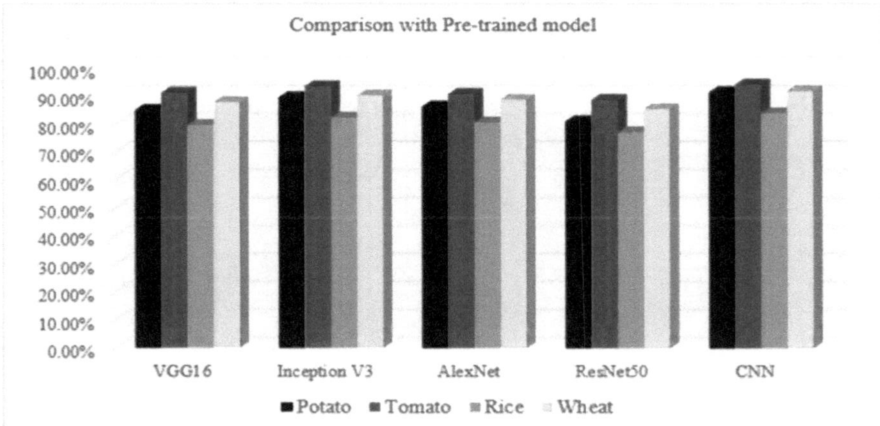

FIGURE 4.4 (a) Comparison of Machine Learning Models. (b) Comparison with Pre-Trained Models. (c) Accuracy and Loss of CNN with Soft-Attention and Inception ResNetV2 Model.

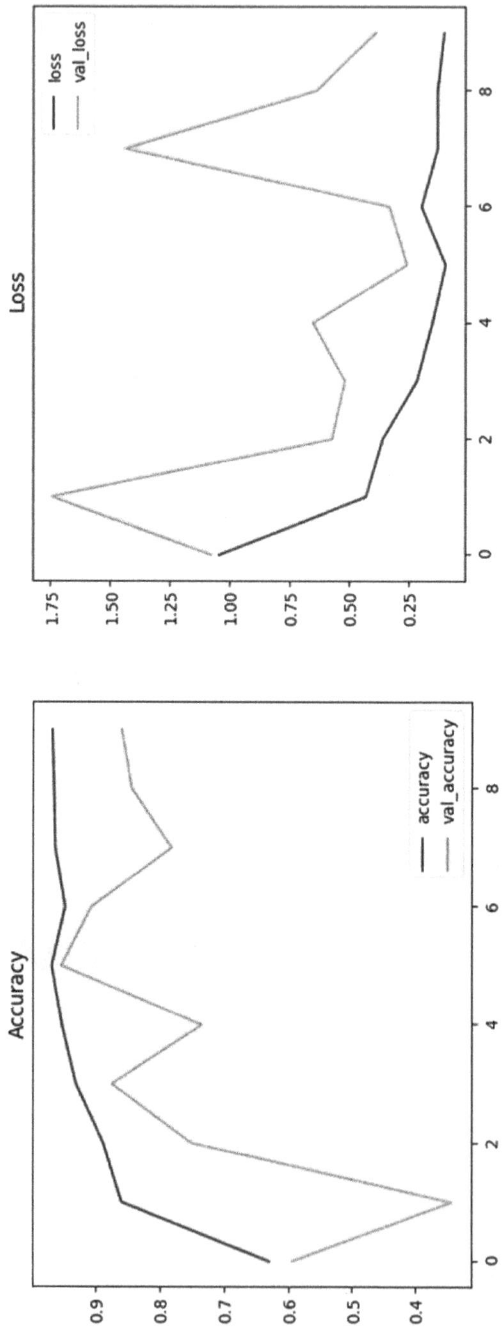

FIGURE 4.4 (Continued)

4.8 CONCLUSION AND FUTURE DIRECTIONS

"Applying machine learning to plant disease detection is transforming precision agriculture by enabling early, precise, and effective identification of plant diseases. This is achieved through various models, including Support Vector Machines (SVM), Convolutional Neural Networks (CNNs), Decision Trees, Random Forests, and Naive Bayes Classifiers." This early detection is crucial for preventing disease escalation, protecting crop health and yields, safeguarding farmers' incomes, and reducing pesticide use to preserve ecosystems. These advancements enhance agricultural productivity and sustainability, allowing for targeted interventions which are key aspects of precision agriculture. Future research should refine these models, develop algorithms for diverse conditions, and ensure transparency and accountability, ultimately promoting economic stability and environmental sustainability in agriculture.

ACKNOWLEDGMENT

Financial support for this study is funded by the Council of Science & Technology U.P. Vigyan Bhawan, 9, Nabiullah Road, Lucknow-226018.

REFERENCES

1. Padol PB and Yadav AA. 2016. "SVM classifier based grape leaf disease detection." *Conference on Advances in Signal Processing (CASP)* (pp. 175–9). IEEE.
2. Agarwal M, Singh A, Arjaria S, Sinha A, and Gupta S. 2020. "ToLeD: Tomato leaf disease detection using convolution neural network." *Procedia Computer Science. 163*, 293–301.
3. Brahimi M, Boukhalfa K, and Moussaoui A. 2017. "Deep learning for tomato diseases: Classification and symptoms visualization." *Applied Artificial Intelligence.* 299–315.
4. Cruz AC, Luvisi A, De Bellis L, and Ampatzidis Y. 2017. "Vision-based plant disease detection system using transfer and deep learning." In 2017*Asabe Annual International Meeting* (p. 1). American Society of Agricultural and Biological Engineers.
5. Friedman N, Geiger D, and Goldszmidt M. 1997. "Bayesian network classifiers." *Machine Learning.* 131–63.
6. Fuentes A, Yoon S, Kim SC, and Park DS. 2017. "A robust deep-learning-based detector for real-time tomato plant diseases and pests recognition." *Sensors 17*(9): 2022.
7. Sokolova M and Lapalme G. 2009. "A systematic analysis of performance measures for classification tasks." *Information Processing & Management.* 427–37.
8. Gandhi R, Nimbalkar S, Yelamanchili N, and Ponkshe S. 2018. "Plant disease detection using CNNs and GANs as an augmentative approach." *IEEE International Conference on Innovative Research and Development (ICIRD).* 1–5.
9. Barnwal, MK, Kotasthane, A, Magculia, N, and Mukherjee, PK, and Agrios GN (2005) *Plant Pathology*, Elsevier Academic Press, San Diego, CA. Alfano, JR and Collmer A (1996) "Bacterial Pathogens in Plants: Life up against the Wall." *Plant Cell. 88*: 1683–98. Allorent, D and Savary S (2005) "Epidemiological characteristics of angular leaf spot of bean: a systems analysis." *European Journal of Plant Pathology 113*: 329–341. *Modeling the Interactions between Host Dynamics and Epidemics of Foliar Diseases in Three Plant Pathosystems*, *88*, 137.

10. Hastie T, Tibshirani R, and Friedman J. 2009. *The Elements of Statistical Learning: Data Mining, Inference, and Prediction.* Springer.
11. Hossain S, Mou RM, Hasan MM, Chakraborty S, and Razzak MA. 2018. "Recognition and detection of tea leaf's diseases using support vector machine." *14th International Colloquium on Signal Processing & Its Applications (CSPA).* 150–4.
12. Huang T, Yang R, Huang W, Huang Y, and Qiao X. 2018. "Detecting sugarcane borer diseases using support vector machine." *Information Processing in Agriculture.* 74–82.
13. Islam M, Dinh A, Wahid K, and Bhowmik P. 2017. "Detection of potato diseases using image segmentation and multiclass support vector machine." *30th Canadian Conference on Electrical and Computer Engineering (CCECE).* 1–4.
14. James G, Witten D, Hastie T, and Tibshirani R. 2013. *An introduction to statistical learning* (vol. 2, p. 18). New York: Springer.
15. Barbedo JGA. 2016. "A review on the main challenges in automatic plant disease identification based on visible range images." *Biosystems Engineering, 144:* 52–60.
16. Quinlan JR. 1986. "Induction of decision trees." *Machine Learning.* 81–106.
17. Kaur S, Pandey S, and Goel S. 2018. "Semi-automatic leaf disease detection and classification system for soybean culture." *IET Image Processing.* 1038–48.
18. Korada NK, Kuma N, and Deekshitulu YVNH. 2012. "Implementation of naïve Bayesian classifier and ada-boost algorithm using maize expert system." *International Journal of Information Sciences and Techniques (IJIST),*
19. Lee SH, Chan CS, Wilkin P, and Remagnino P. 2015. "Deep plant: Plant identification with convolutional neural networks." *International conference on image processing (ICIP).* IEEE. 452–6.
20. Li L, Zhang S, and Wang B. 2021. "Plant disease detection and classification by deep learning—A review." *IEEE Access: Practical Innovations, Open Solutions.* 56683–98.
21. Ramakrishnan M. 2015. "Groundnut leaf disease detection and classification by using back propagation algorithm." In: *International Conference on Communications and Signal Processing (ICCSP).* IEEE. 964–68.
22. Majumdar D, Ghosh A, Kole DK, Chakraborty A, and Majumder DD. 2015. "Application of fuzzy C-means clustering method to classify wheat leaf images based on the presence of rust disease." *Proceedings of the 3rd International Conference on Frontiers of Intelligent Computing: Theory and Applications (FICTA) 2014.* Springer. 277–84.
23. Masazhar ANI and Kamal MM. 2017. "Digital image processing technique for palm oil leaf disease detection using multiclass SVM classifier." *4th International Conference on Smart Instrumentation, Measurement, and Application (ICSIMA).* 1–6.
24. Mohanty SP, Hughes DP, and Salathé M. 2016. "Using deep learning for image-based plant disease detection." *Frontiers in Plant Science, 7,* 1419.
25. Munjal D, Singh L, Pandey M, and Lakra S. 2023. "A systematic review on the detection and classification of plant diseases using machine learning." *International Journal of Software Innovation (IJSI).* 1–25.
26. Owomugisha G, Melchert F, Mwebaze E, Quinn JA, and Biehl M. 2018. "Machine Learning for diagnosis of disease in plants using spectral data." *International Conference on Artificial Intelligence (ICAI).* 9–15.
27. Paulus S, Behmann J, Mahlein AK, Plümer L, and Kuhlmann H. 2014. "Low-cost 3D systems: Suitable tools for plant phenotyping." *Sensors (Basel).* 3001–18.
28. Singh AK, Sreenivasan SVN, Mahalaxmi USBK, Sharma H, Patil DD, and Asenso E. 2022. "Hybrid feature-based disease detection in plant leaf using convolutional neural network, Bayesian optimized SVM, and random forest classifier." *Journal of Food Quality, 2022*(1): 2845320.

29. Soman KP, Loganathan R, and Ajay V. 2009. *Machine Learning with SVM and Other Kernel Methods*. PHI Learning Pvt Ltd.
30. Sujatha R, Chatterjee JM, Jhanjhi NZ, and Brohi SN. 2021. "Performance of deep learning vs machine learning in plant leaf." *Microprocessors and Microsystems, 80*, 103615.
31. Fawcett T. 2006. "An introduction to ROC analysis." *Pattern Recognition Letters*. 861–74.
32. Umapathy Eaganathan D, Lackose V, and Benjamin FJ. 2014. "Identification of sugarcane leaf scorch disease using K-means clustering segmentation and K-NN based classification." *International Journal of Advances in Computer Science and Technology (IJACST)*. 11–6.
33. Govardhan M and Veena MB. 2019. "Diagnosis of tomato plant diseases using random forest." *2019 Global Conference for Advancement in Technology (GCAT)*. 1–5.

5 IoT and Machine Learning-Based Controlled Drip Irrigation and Disease Detection in Perlette Grape Cultivation Using Sensor and Image Technology

M. Sucharitha and Ravindrakumar Selvaraj

5.1 INTRODUCTION

In peninsular India due to its sub-tropical climate, perlette Grape is cultivated and became an important crop. In India, grapes are cultivated in Maharashtra, Bangalore, Kolar, Bijapur, followed by Tamil Nadu and Andhra Pradesh. Precision irrigation is the only solution to improve the growth of grapes. The method controls the quantity of nutrient and water to be supplied to the vine growth. Perlette Grape climbing plant cultivated in larger part of Tamil Nadu and Andhra Pradesh. The problems due to hail, downy mildew, frost injury on buds, powdery mildew, mature sporulating lesions and lesions after sporulation are to be controlled. Traditional cultivation methods often rely on manual monitoring and intervention, which can be time-consuming and inefficient. Overexploitation of this plant occurs due to insufficient scientific knowledge of the local people. Perlette Grape is a high-value crop with significant economic potential, but its cultivation is challenging due to the plant's susceptibility to fungal infections and the need for precise irrigation management. Fungal infection and cultivation were investigated. It has been found that there were various types of grape variety in and around the world, namely Anab-e-Shahi, Banquiabyad Foster's & beauty seedling, Pusa Seedless, Thompson Seedless, Perlette, Kairon, Kandhari, Khalili, Pandhari Sahebi, Early Muscat, Black Muscat and Cardinal. Artificial intelligence (AI) and machine learning (ML) will be used

 DOI: 10.1201/9781003487005-5

to monitor the crop growth and controlled irrigation through feedback system. The field can be monitored using Internet of Things (IoT). The product/research work will involve the IoT system design and development. Inputs are from the humidity, soil moisture sensors and temperature. The water requirements were predicted from the sensor inputs given to the machine learning algorithm. The early signs of fungal infection can be predicted using the predetermined template. The post-harvest losses can be avoided using ML algorithms towards plant monitoring, irrigation and fungal infection detection. Recommended Dose of Fertilizer (RDF) is applied through drip irrigation. Image technology using ML will be utilized to monitor the fungal infection, plant height and flower count using AI and ML. The system will be deployed in a test plot where Perlette Grape is being cultivated. The system will monitor and regulate the irrigation process, and in cases of fungal infection, alert the farmer to take appropriate action. The system will also provide real-time data on the plant's growth and health, enabling farmers to make informed decisions about crop management. The expected outcomes of the work are improved crop yield, reduced water consumption and improved plant health. The system will help farmers optimize the use of resources and improve their profitability.

5.2 LITERATURE REVIEW

Several research articles, field reports, agriculture data and testimonies were collected supporting the work. Material and journals based on Machine Learning, Internet of things, Irrigation, Fungal Infection and Cultivation were investigated. In past years several machine learning methods were used for monitoring the performance of various units of irrigation system [1–2]. The machine plays roles in other areas like wind prediction, water management, irrigation and evapotranspiration analysis [3–5]. Artificial Neural Network were used in dripper water distribution evaluation. An optimistic approach is made to control the water distribution via the dripper units [6]. For humid regions artificial neural networks were used for estimating reference evapotranspiration [7]. There were challenges in analysing the evapotranspiration with limited data due to accuracy problems. Fuzzy and decision tree algorithms were applied on data for estimating land suitability and irrigation [8–9]. In [10–11] a neural fuzzy based water conservation system for irrigation was presented. The insufficiency of binary based fuzzy models is less efficient compared to other deep learning models [12]. IoT based soil monitoring finds a good solution due to remote monitoring and intelligent scheduling methods. Open-source platforms or different technology were used. Another artificial intelligence method is the decision tree for prediction of an irrigation event occurrence [13–14]. The neural network was part of soil moisture detection, prediction and other parameters [15–18]. To enhance the machine learning decision, a large database is to be recorded from sensors. Limited data will challenge the accuracy. IoT and precision agriculture use statistical and analytical models to predict the plant growth and crop production [19]. The other irrigation system used in recent times is the clustered irrigation system. The crop growth and yield are affected by the fungal infections. Proper precision irrigation with controlled water and nutrient supply can enhance the yield. The issues happening in the grape plant are due to hail, downy mildew, frost injury on buds, powdery mildew, mature sporulating lesions and lesions after sporulation. A Machine Learning system for grape flower analysis and cultivation was done [20–21]. Frequent monitoring through proximal and remote sensing methods were undergone. Spectral vegetation indices (VIs) were analysed. The literature presents an Auto machine learning (AutoML) method and segmentation for grape bunches detection was done [22–23].

5.3 MACHINE LEARNING BASED METHODS

The machine learning methods replace the manually evaluated system. The linear and nonlinear regression machine learning methods were applied for the detection process in agricultural application. Several models like a linear regression model and Yolo [24–25] were applied for the grape detection. The interpretation and non-parametric learning approach of the decision tree method is used in the farming. The decision is to be made on the data interpretation for disease identification also [25]. For other ensemble-based decision tree methods like extra trees, random forests and AdaBoosting, an Optimization algorithm for accurate decision making can be performed by designing and developing hybrid machine learning algorithms. The decision tree is made first followed by the reduction of bias and variance using boosting or bootstrap aggregation. A few ensemble learning methods such as AdaBoost manage imprecise regression using retrains. In random Forest, a supervised learning approach was used through regression by ensemble learning. Here a single model combines multiple decision tree regressors. Extremely Randomized Trees combine predictions from different decision trees with original inputs. The computationally more expensive nature of decision tree implementation is a major disadvantage. Support Vector Machines decide between the subset of the training data for farming analysis.

5.3.1 Machine Learning for Precision Viticulture

To optimize the yield and costs, the Precision viticulture methods were adopted. The acquisition of geolocalized data is used for the crop yield. Early disease detection is an important task. Human based operations are prone to mistakes. Partial automations were available to enhance the process. This integrated method increases the crop production using precise data and control. The precision viticulture tools were:

GPS, Global Position System for determination of the location and position of the field. Site location is noted.

GIS, Global Information System, the characteristics of the field studied using the geodata. The geostatistical algorithms is implemented based on maps which give the field or site information.

GNSS Global Navigation Satellite System provides proper control of vehicles used in field like the robots.

VRT, Variable Rate Technology, is applied to machines which does the soil management, weed removal and fertilizer control.

Multispectral and hyperspectral cameras capture images which are used to detect the details of chlorophyll content and crop stress level.

Sensors measure the crop, operation and environmental. UAV, Unmanned Aircraft System makes the image possible when taken from high altitude.

5.3.2 Irrigation Methods

When considering the irrigation method, grape vine root growth patterns are to be investigated for an efficient cultivation. In India during drought conditions, subsurface drip-irrigated crops may survive the heat when the root system is confined. Using the soil core methods, the water requirement and the problems due to the deficit can be evaluated. The deficit may affect the root growth which is identified. The other parameter to be measured is soil moisture variations. Due to climatic conditions and environmental factors fluctuations in the moisture happens. The other challenges are due to the adaptability of soil properties, sensor placement, sensor characteristics and plot location. The values are to be calibrated before acquisition of exact real-time values. The climate conditions are to be evaluated based on the data obtained from the sensor units. The unexpected or unpredicted climatic changes like heavy rain, heat or dry season and excessive precipitation will affect the yield. The most important phenomenon by which grapevines effectively manage drought stress is through hydraulic conductivity.

Environmental fluctuations have affected the Mediterranean region, which is the producer of 50% of the world's wine grapes. From several field works it has been observed that the change in climate drastically affects the biochemical attributes. In addition the physiological characteristics of the grape plants also change. But it has been very critical to consider the times during early stomatal closure for preserving foliar water. The water transportation to leaves from the roots is managed by the channel proteins. Cellular hydraulic conductivity is altered for permeability control and problems due to water transportation. On the other hand, high tension happens through transpiration which removes water from the leaf. Biotic and abiotic stress factors like drought and phylloxera tolerance are required by the rootstocks.

5.4 PROPOSED DESIGN

5.4.1 Experimental Setup for Implementing the Proposed Methodology

5.4.1.1 Field Preparation

The field is prepared for the cultivation of the plant. The ploughing and removal of gravel, stones are performed using manual labour. Three thousand square feet of land is chosen for the system to be designed with a distributed irrigation unit. The main parts of the field preparation are:

a. Size of the field – 3000 square feet
b. 'V' shaped tubers – 110 g each were selected, treated with Bavistin at 0.1% for 30 minutes and sown in raised beds with a spacing of 1.82 m between rows and 12.5 cm within rows

c. Drip Irrigation Component [Randomized Block Design (RBD) and Cumulative Pan Evaporation (CPE)]
d. Fertilizer unit
e. Nutrient flow control 1 l/minute

5.4.1.2 Solution Preparation/Nutrient Mix Preparation

These nutrients are becoming part of the irrigation unit. The nutrients are supplied through the drip irrigation. The most popular existing drip irrigation system is shown in Figure 5.1. The water pump is where the proposed method is highly designed (Figure 5.2). The other units like filter unit, valves, pressure gauge and

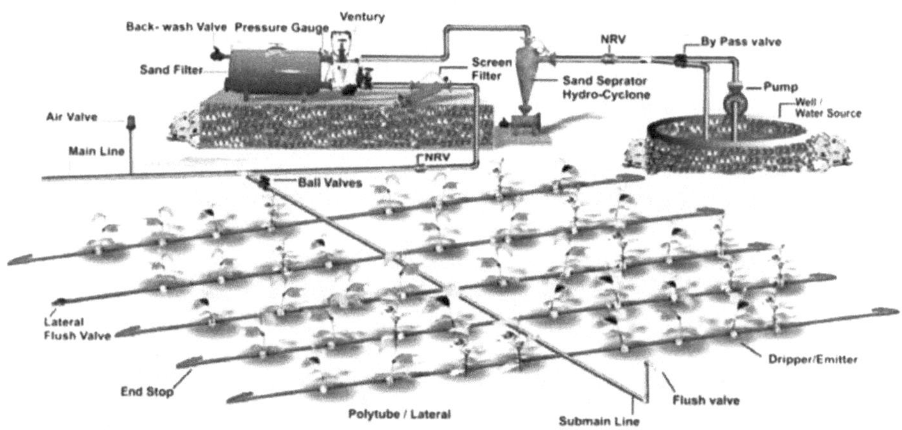

FIGURE 5.1 Conventional drip irrigation system.

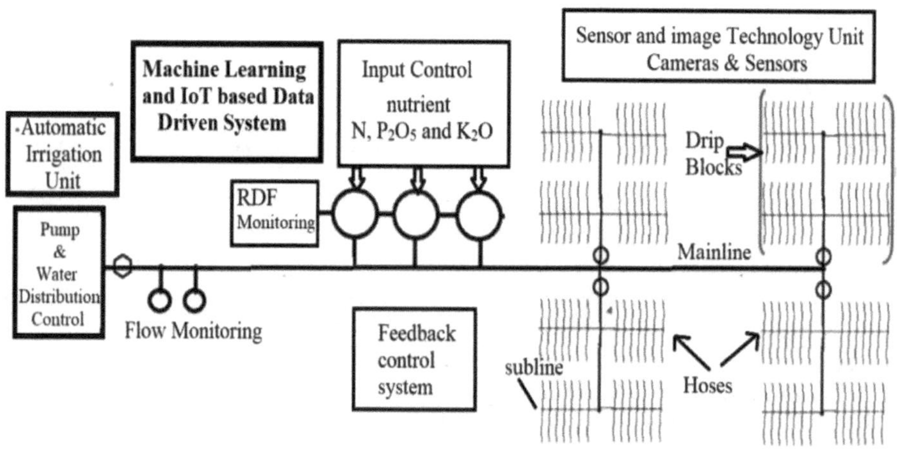

FIGURE 5.2 Block diagram of the proposed Perlette irrigation method.

fertilizer unit were controlled by machine learning and the data are acquired by IoT sensors.

5.4.1.3 Hardware Design

a. Microcontroller kit: Hardware platform will be the Arduino and Raspberry kit
b. Water Temperature DS18B20 monitoring. The resolution is variable to 9, 10, 11 or 12 bits,
c. Air Temperature- DHT11 Temperature sensor fan could be controlled if the air temperature is too high
d. Humidity monitoring/soil moisture
f. DS3234 Real Time Clock
g. Data acquisition – The data acquisition unit does the signal conditioning using the sensors and ADC.
h. Solar power panel and charging circuit with battery.

5.4.1.4 Monitoring Unit

This monitors the following through the web/mobile application like Soil Moisture: air temperature, water and fertilizer reservoir level, pH status and pest. Automated monitoring and controlling environmental parameters like soil moisture, temperature, humidity, PH, conductivity made possible with the Internet of Things. A smart solar power unit will reduce the cost of conventional electrical energy. The low cost, portable prototype device for monitoring the irrigation, pest and fungal infection were developed. The embedded device with IoT monitored the field parameters and performs autonomous control. The variations in the soil moisture are also monitored. The flow diagram for the proposed perlette irrigation system is shown (Figure 5.3). The data collected from the multiple sensors and units is processed through the data processing unit of the embedded system.

The feature extraction uses wavelet transform and Non-negative Matrix Factorisation (NMF) techniques. The extracted features are analysed using the Machine Learning classification unit where decision-making is done for control of pumps, pesticide sprayer, nutrient distribution, water distribution, solar power monitoring. The ML unit uses Convolution Neural Network architecture combined with optimization technique. The dataset is taken from the agricultural research laboratories and the training and testing datasets are made. The decision-making controls the control unit setup which has several electronic and electrical units to run the motors, pumps etc. Sensor arrays monitor the environment, temperature and humidity.

The AI unit always has sensor units which have to be investigated, monitored and analysed. These data are the controlling element of the system. The sensors are placed at equidistance so the measurement is very perfect. The data is in analog which has to be converted to digital. Digital units are the need of the day. These units process only the digital values which have to be converted from analog. The voltage converters and amplifiers help in boosting the input amplitude so the required data is better. The pH

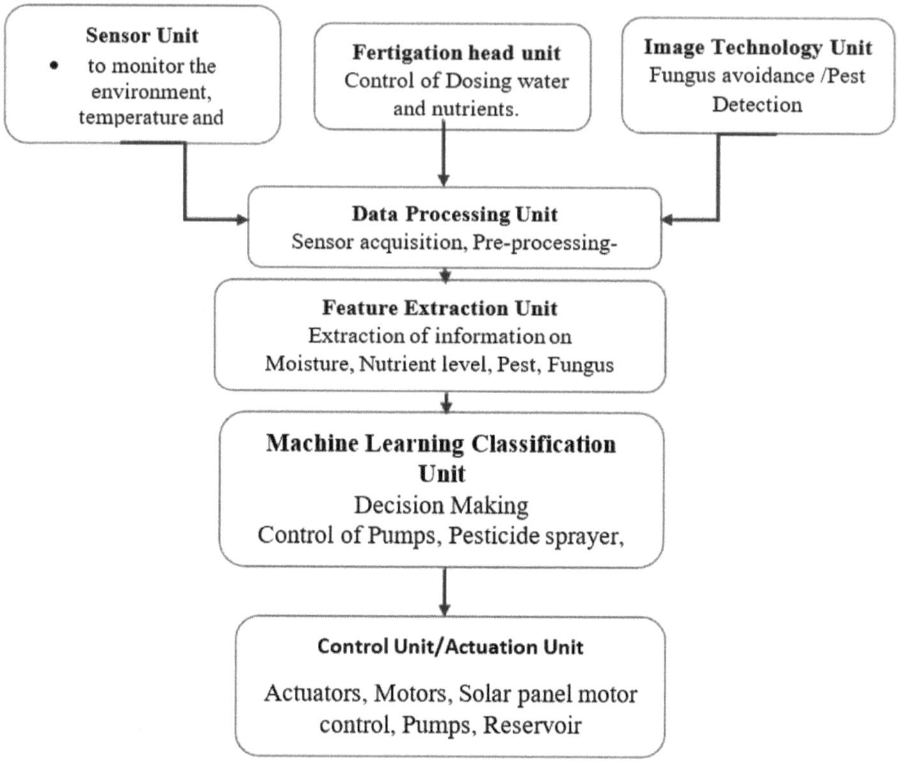

FIGURE 5.3 Process flow diagram.

and soil moisture are monitored using the respective sensors. This setup also uses the analog to digital unit followed by the microcontrollers.

1. Fertigation head unit, which enables accurate dosing of nutrients and drip irrigation module.
2. IoT-embedded based design – Raspberry kit, domain webpage etc.
3. Design of ML algorithm.
4. Solar Based Power – Solar cell, batteries, charging circuit and other components.
5. GSM module for Transmission and reception – GSM unit, SIM, embedded interfaces.
6. Nutrition Level Monitoring – nutrient solution with solid fertilizer salts, containers, sensors, pumps, relays.
7. Sensor Interfacings – relays, amplifiers, filters.
8. Microcontroller based autonomous control Arduino (optional).
9. Feedback control through web and mobile based design.
10. Pest control unit.

5.4.1.4.1 *Key Units in the Design*

- Sensor and image technology for the detection of plant pathology and to control fungus growth. The fungus growth affects productivity and it spreads quickly. Initial stages of diagnosis will help in the increase of production. Drone based detection is suitable for larger fields. The implementation of swarm intelligence to coordinate the process improves the diagnosis steps. Monitoring the environmental parameters like soil moisture, temperature, humidity and Ph. The data processing units are designed using LIC circuits and ADC units. The output of ADC is connected with the microcontrollers.
- ML-IoT based autonomous plant monitoring system to monitor moisture, air, nutrients, temperature and sunlight. The nutrient overdose sometimes shows if a fungus has affected the leaves, so the monitoring unit should keep track of the fungus problem and nutrient given.

The hardware components used are Raspberry pi, sensors, pumps. The software development as embedded C, python, HTML, PHP.

- Feedback control system for irrigation and fertilization to increase the crop production.
  Control units are designed to drive the pumps for water and fertilizer. The design also monitors if any deficiency occurred in the plant.
- Continuous monitoring of plant growth and data analysis using ML.
  Sensors and other monitoring equipment are used to collect data on plant growth and the data is analysed using an ML algorithm to make predictions.

5.4.1.4.2 *Machine Learning for Grape Disease Detection*

Early machine learning methods were used in agriculture. In recent times, deep learning methods play an important role in all fields. Apart from the conventional methods based on fuzzy logic, Artificial neural networks (ANN), expert systems, IoT based design enhances the way machine learning works. The machine learning based methods when combined with IoT enhance the control elements. In certain applications, semi-automated methods have grown replacing the traditional agriculture approach. Classification of diseases in grape plant like Black rot, Leaf blight, Black Measles (esca), Powdery mildew, is implemented. The investigation of machine learning methods performed were Support vector machine classification, CNN – VGG16, CNN, Random Forest classification. The control system is operated by the IoT module, the hardware configuration was already presented in previous sections. The surrounding conditions and irrigation process are initiated based on the classification output of the machine learning system. Predetermined agriculture data and predictable results are scrutinized for control mechanism. The method achieves the objectives mentioned previously based on the data obtained from the IoT modules.

5.4.1.4.3 IoT Technology in the Vineyard

The key environmental data are collected and compared with the preassigned data for control operation by the machine learning method. The data respective to micro-climatic condition, soil, and vine physical conditions is collected. The other parameters monitored through the combination of ML and IoT is the water stress, vine health status, pruning period, optimum time for harvesting and to optimize crop load. The other most important problem addressed is the determination of grapevine diseases like mildews and botrytis. Previous analytics and observed IoT data are compared and classified using ML

5.4.1.4.4 Data Analytics in Vineyard Data

The machine learning based methods analyse the Big data (BD) which were machine-readable. The data were digital, raw, unstructured. The data which is the large database of the grape field contains the data of the field, climatic conditions, irrigation information, crop details, growth information, disease diagnosis, fertilizer or insecticide usage etc. This data is collected on daily basis for several years to have a large database.

The machine learning methods are programmed in the IoT hardware to learn and problem-solve and take necessary actions. Nowadays Deep learning (DL) methods were highly used in large computation and BD.

The data collected will be based on the field, crop, climate and these BD will be used by AI. Table 5.1 shows the data details for parameters. There may be more based on the field and other features.

The Big data collected should contain the information about the above details intuitively. The Agri-Food Experiment Ontology (AFEO) forum have analysed and given guidelines to merge the technology to be combined for the purpose of better viticulture practices, wine-making products, and operations. The Big data, IoT and machine learning methods analyzing field data are presented in Figure 5.4.

Data acquisition/Big data are unstructured data which is acquired from the sensors, geological data, and field information. These data can be structured and applied to the data processing unit where missing data is filled. Preprocessing and analyses using cumulative distribution is done. These are given as input to the machine learning approach. Next, the vineyard prediction and forecast is done.

Some of the classification performed are

- Disease and pest control
- Suitability requirements
- Grape/grape bunch detection
- Biotic factors and stress
- Grapevine phenology detection, yield prediction
- Wine quality, classification
- Traceability, authenticity, protection

TABLE 5.1
Common data analysis with anticipated outcomes

Data details	Inclusive information for AI based analysis	Anticipated outcomes/ Implications
Vineyard Landscape	• Latitude • Elevation • Soil structure • Tillage • Water capacity • Weeds • Disease problems	• Weeds • Diseases • Climate change • Crop production • Pests
Meso-climate	• Ambient temperature–Day • Ambient temperature–Night • Solar radiation • Precipitation • Humidity • Evaporation	• Irrigation scheduling • Nutrient management • Canopy management • Beneficiary insects • Waste management
Micro climate	• Soil moisture • Leaf wetness • Vine physiology • Fruit composition	• Precision agriculture • Phenology of grape • Smoke taint

These combine impact production, blending, aroma, abnormal fermentation, market.

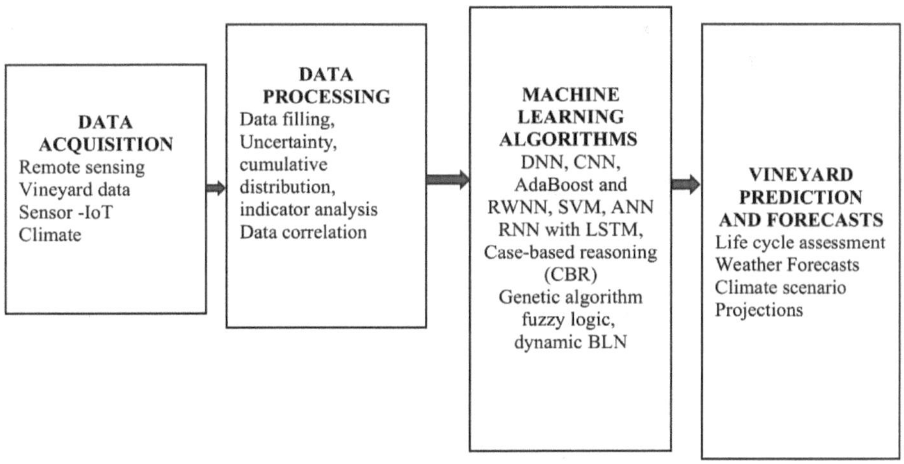

FIGURE 5.4 Block diagram of big data, machine learning and forecast relations.

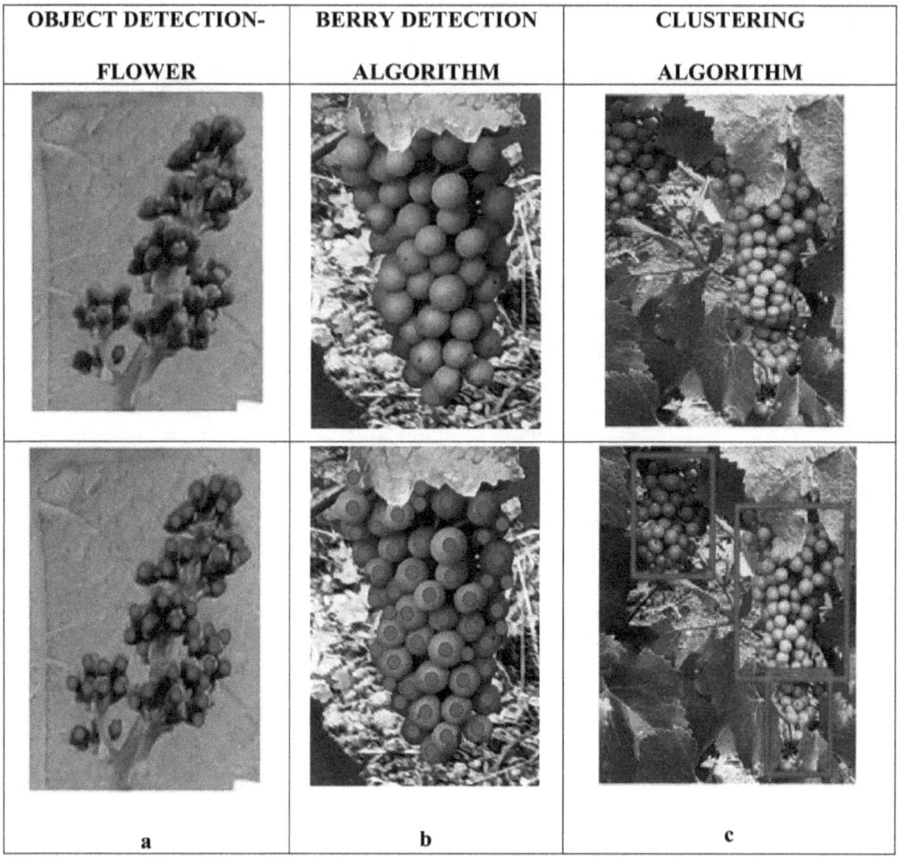

FIGURE 5.5 Application of ML in field.

Evaluation of machine learning and the deep learning model is shown in Figure 5.5a; for object detection, different varieties of grape plant images were considered, methods like FCN, H-maxima transform and generic keypoint algorithm were used.

In Figure 5.5b, for berry detection algorithm, Mahalanobis segmentation and counting with a Boolean model and FCN model were used. The other algorithms which can be done through 3D construction are the SVD, SVM and R-CNN. In Figure 5.5c, grape detection through clustering is performed using the Active contour segmentation, Color-based thresholding, Color features and AdaBoost and Spatial Clustering. Deep learning methods were used for the grape detection.

5.4.1.4.5 Machine Learning-Based Grape Leaf Disease Diagnosis

The diagnosis of grape leaf diseases consists of the background removal, segmentation feature extraction, selection of features and classification. Some of the diseases

Leaf Blight- Tan/dark brown spots with dark borders	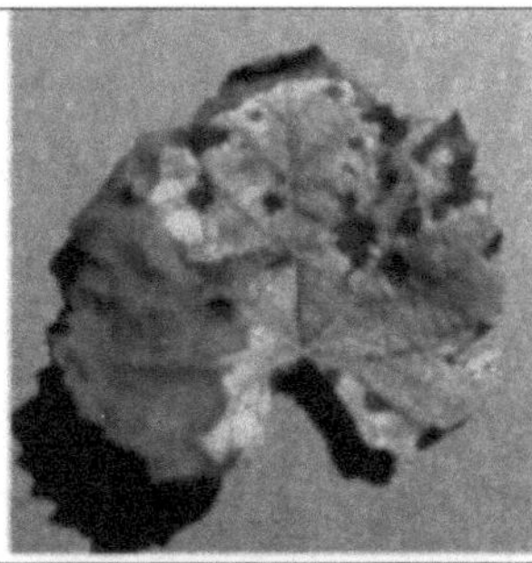
Black Rot-Brown circular lesions	
Black Measles-Tiger stipe pattern	
Features Type 1	gray-level co-occurrence matrix (GLCM) binary robust invariant scalable key-points (BRISK) histogram of oriented gradients (HOG) local binary patterns (LBP) oriented fast and rotated BRIEF (ORB)
General features	Contrast, energy, homogeneity, mean, standard deviation, entropy, root mean square, variance, smoothness, kurtosis, Skewness,

FIGURE 5.6 Diseases affecting the grape plant and its features.

affecting the grape plant are shown in Figure 5.6. The figure shows the different diseases found in grape plant and the features are given at the bottom of the figure. In recent times in several applications the deep learning method, CNN-based architectures, like GoogLeNet, AlexNet, GoogLeNet, were used.

5.5 CONCLUSION

In this work, the Perlette Grape cultivation controlled irrigation and protection from fungal infections were discussed in detail. The problems due to diseases and methods were presented. Optimizing the cultivation process by controlled drip irrigation and disease detection processes using sensor-based IoT devices and advanced machine learning algorithms were presented. In addition, AI technology was used in image classification, plant growth, mineral monitoring and pest control. The work involves the design and development of an IoT system for soil and field monitoring using sensors. The control units for the autonomous irrigation are managed by the decision of the machine learning algorithm. The nutrient and water requirements were predicted in advance based on the early signs of fungal infection. In future the system will be deployed in a test plot where Perlette Grape is being cultivated. The system will monitor and regulate the irrigation process, and in cases of fungal infection, alert the farmer to take appropriate action. The system will also provide real-time data on the plant's growth and health, enabling farmers to take necessary action in advance.

REFERENCES

1. Al-Ghobari, H.M., El-Marazky, M.S., Dewidar, A.Z., and Mattar, M.A. (2018). Prediction of wind drift and evaporation losses from sprinkler irrigation using neural network and multiple regression techniques. Agricultural Water Management 195: 211–221.
2. Awal, R., Habibi, H., Fares, A., and Deb, S. (2020). Estimating reference crop evapotranspiration under limited climate data in West Texas. Journal of Hydrology: Regional Studies 28: 100677.
3. Blasi, A.H., Abbadi, M.A., and Al-Huweimel, R. (2021). Machine learning approach for an automatic irrigation system in Southern Jordan Valley. Engineering, Technology & Applied Science Research 11 (1): 6609–6613.
4. Dias, S.H.B., Filgueiras, R., Filho, E.I.F., Arcanjo, G.S., Da Silva, G.H., Mantovani, E.C., and Da Cunha, F.F. (2021). Reference evapotranspiration of Brazil modeled with machine learning techniques and remote sensing. PLoS ONE 16 (2): 1–20 .
5. Feng, Y., Cui, N., Zhao, L., Hu, X., Gong, D (2016). Comparison of ELM, GANN, WNN and empirical models for estimating reference evapotranspiration in humid region of Southwest China. Journal of Hydrology 536: 376–383.
6. Ferreira, L.B., Duarte, A.B., Da Cunha, F.F., and Fernandes Filho, E.I (2019). Multivariate adaptive regression splines (MARS) applied to daily reference evapotranspiration modelling with limited weather data. Acta Scientiarum. Agronomy 41 (1): 39880.
7. Filgueiras, R., Almeida, T.S., Mantovani, E.C., Dias, S.H.B., Fernandes-Filho, E.I., Da Cunha, F.F., and Venancio L.P. (2020). Soil water content and actual evapotranspiration predictions using regression algorithms and remote sensing data. Agricultural Water Management 241: 106346.
8. Hoseini, Y. (2019). Use fuzzy interface systems to optimize land suitability evaluation for surface and trickle irrigation. Information Processing in Agriculture 6 (1): 11–19.

9. González Perea, R., Camacho Poyato, E., Montesinos, P., and Rodríguez Díaz J.A (2019) Prediction of irrigation event occurrence at farm level using optimal decision trees. Computers and Electronics in Agriculture 157: 173–180.
10. Gu, Z., Zhu, T., Jiao, X., Xu, J., and Qi, Z (2021). Neural network soil moisture model for irrigation scheduling. Computers and Electronics in Agriculture 180 (1): 105801.
11. Kelley, J., and Pardyjak, E (2019). Using neural networks to estimate site-specific crop evapotranspiration with low-cost sensors . Agronomy 9 (2): 108.
12. Jamroen, C., Komkum, P., Fongkerd, C., and Krongpha, W (2020). An intelligent irrigation scheduling system using low-cost wireless sensor network toward sustainable and precision agriculture. IEEE Access 8: 172756–172769.
13. Keswani, B., Mohapatra, A.G., Mohanty, A., Khanna, A., Rodrigues, J.J.P.C., Gupta, D., and De Albuquerque, V.H.C. (2019). Adapting weather conditions based IoT enabled smart irrigation technique in precision agriculture mechanisms. Neural Computing and Applications 31: 277–292.
14. Kontogiannis, S., Kokkonis, G., Ellinidou, S., and Valsamidis, S (2017) Proposed Fuzzy-NN algorithm with LoRaCommunication protocol for clustered irrigation systems. Future Internet 9 (4): 78.
15. Mouatadid, S., Adamowski, J.F., Tiwari, M.K., and Quilty, J.M (2019) Coupling the maximum overlap discrete wavelet transform and long short-term memory networks for irrigation flow forecasting . Agricultural Water Management 219: 72–85.
16. Munir, M.S., Bajwa, I.S., Naeem, M.A., and Ramzan, B. (2018). Design and implementation of an IoT system for smart energy consumption and smart irrigation in tunnel farming. Energies 11 (12): 3427.
17. Seyedzadeh, A., Maroufpoor, S., Maroufpoor, E., Shiri, J., Bozorg-Haddad, O., and Gavazi, F. (2020). Artificial intelligence approach to estimate discharge of drip tape irrigation based on temperature and pressure. Agricultural Water Management 228: 105905.
18. Torres-Sanchez, R., Navarro-Hellin, H., Guillamon-Frutos, A., San-Segundo, R., Ruiz-Abell6n, M.C., and Domingo-Miguel, R (2020). A decision support system for irrigation management: Analysis and implementation of different learning techniques. Water 12 (2): 548.
19. Torres, A.B.B., Da Rocha, A.R., Coelho da Silva, T.L., De Souza, J.N., and Gondim, R.S (2020) Multilevel data fusion for the internet of things in smart agriculture. Computers and Electronics in Agriculture 171: 105309.
20. Tello, J., Herzog, K., Rist, F., This, P., and Doligez, A. (2019). Automatic flower number evaluation in grapevine inflorescences using RGB images. American Journal of Enology and Viticulture 71: 10–16.
21. Palacios, F., Bueno, G., Salido, J., Diago, M.P., Hernández, I., and Tardaguila, J. (2020). Automated grapevine flower detection and quantification method based on computer vision and deep learning from on-the-go imaging using a mobile sensing platform under field conditions. Computers and Electronics in Agriculture 178: 105796.
22. Ghiani, L., Sassu, A., Palumbo, F., Mercenaro, L., and Gambella, F. 2021. In-field automatic detection of grape bunches under a totally uncontrolled environment. Sensors 21: 3908.
23. Chen, S., Song, Y., Su, J., Fang, Y., Shen, L., Mi, Z., and Su, B. (2021). Segmentation of field grape bunches via an improved pyramid scene parsing network. International Journal of Agricultural and Biological Engineering 14:185–194.

24. Sozzi, M., Cantalamessa, S., Cogato, A., Kayad, A., and Marinello, F. (2022). Automatic bunch detection in white grape varieties using YOLOv3, YOLOv4, and YOLOv5 deep learning algorithms. Agronomy 12: 319.
25. Too, E.C., Yujian, L., Njuki, S., and Yingchun L. (2019). A comparative study of fine-tuning deep learning models for plant disease identification. Computers and Electronics in Agriculture 161: 272–279.

6 Towards Precision Agriculture

In-Situ Automated Greenhouse Monitoring and Controlling Using IoT and Machine Learning

Kaviya P, Selvakumar B, Ganga A, and Kavitha S

6.1 INTRODUCTION

For most Indian farmers, agriculture is their primary source of income, and it has a large impact on the country's economic development. In the current scenario, unanticipated climatic changes have an impact on crop productivity. The occurrence of imbalanced climatic changes is mainly attributed to anthropogenic activities. As a result, we receive low rainfall, cyclones, floods, and drought due to unusual summer seasons. Farmers struggle to cope up with the variations in weather as there are no technology-assisted solution to combat it. Besides, farmers are led to tackle unforeseen difficulties in yielding high crop production.

Optimum environmental conditions are pivotal for higher productivity. A consistent weather pattern is required to sustain the ideal climatic conditions necessary for producing large yields. Recent advancements in technologies have helped us to achieve such conditions for a longer period of time without the involvement of manpower, except for the overall monitoring of the data received. Advanced technologies such as cloud computing, mobile applications, machine learning, and the IoT are a few examples. This chapter demonstrates a suitable approach that combines the technologies described above for effective agriculture management.

On the grounds of scientific findings, for the issue at the hand, installation of cost-effective *IsAGHMCS* can be a good remedy. *IsAGHMCS* utilizes IoT that has various sensors to monitor humidity, temperature, soil moisture, air quality, CO_2, and soil nutrient levels to analyse the essential parameters that boost crop productivity. ML is one of the versatile technologies that lets us to forecast various parameters with much accuracy. For a farmer who is unaware of high-end technologies and often relies on the support of educated personnel right from soil health prediction to crop yield prediction, ML can provide results in a farmer-understandable format. Hence, a farmer

DOI: 10.1201/9781003487005-6

no longer needs an educated counsellor for farming. A cloud analytics platform is used to store the observed environmental data. The strategic use of scalable cloud storage saves space and time required if the cloud is not used. A mobile application called *"Vivasayi"* is developed to know the live update from the farm in the farmer's vernacular language.

Hence, this approach will be very helpful for farmers to monitor and control their fields effectively. This system seamlessly increases the crop yield using optimal environmental parameters in the field. Besides, automated farming helps to achieve maximum yield while saving time and resources. In addition, *IsAGHMCS* aims to consume a minimum amount of water, fertilizers, and pesticides to produce maximum output by increasing accuracy. Since IoT and ML enhances efficiency, it lessens the operational costs and human folly.

6.2 REVIEW OF LITERATURE

This literature review delves into current research on smart agriculture. In recent studies, many researchers adopted IoT and AI technologies in smart farming [1][2] like field monitoring and controlling [3][4][5], disease detection, yield prediction, and weed management [6]. Smart farming methodologies are discussed as follows:

Aburasain et al. [7] have developed a hybrid deep learning (HDL) with enhanced black widow optimization for a smart farming system. This model employs the bald eagle search (BES), HDL classification model for selecting a feature subset, intrusion detection, and classification respectively. They have used the ToN-IoT and Edge-IIoTset dataset to carry out simulations. Even the hypersensitive parameters can also be predicted using the EBWO algorithms. It is evident that smart farming is achievable with the deployment of suitable IoT devices and training them appropriately.

Astillo et al. [8] suggested security solutions for smart greenhouse farming (SGF). The Proteus Design Suite was used to simulate IoT devices placed in a smart greenhouse setting. Though it needs a small amount of storage space, the method finds errors. ML algorithms such as kNN or SVM were employed to determine which method is optimal for similar applications.

Ayaz et al. [9] conducted a survey on several types of IoT based systems for monitoring and regulating greenhouses. They have done a detailed analysis of the IoT devices and communication protocols linked to wireless sensors utilized in agricultural settings. Also, they have discussed the sensors used to monitor the soil preparation, water irrigation, crop status, pest, and insect detection. They have suggested IoT-based platforms and frameworks for farming.

Pak et al. [10] have created and evaluated path planning algorithms to find the quickest and clearest routes for agricultural harvesting robots. They looked at several automated farming algorithms to highlight the significance of smart farming in reducing crop damage, and the study verified the experimental results. In order to protect crops from harm, the authors emphasized the importance of precise route planning and increased autonomy for autonomous vehicles.

Huang et al. [11] investigated a novel hypothesis: Photovoltaic Agricultural Internet of Things (PAIoT) for achieving sustainable agriculture even when the climate crisis is on the rise. PAIoT assists in automated farmland irrigation but it has limitations. The key concerns are: photovoltaic panel cleaning, node deployment, complete utilization of water resources, cost optimization for IoT devices, node misbehaviour analysis, optimization of data being transported and acquired among the nodes, malicious firm-wares, impact of photovoltaic panel power generation on the ecosystem, curating the functions of photovoltaic modules. The study opens a vast query to be addressed for prospectus smart farming while mitigating the climate crisis.

Ndjuluwa et al. [12] reviewed the technologies and applications of IoT for crop farming. They have focused on sensors to collect the data, communication protocols for transferring the data, cloud/fog computing technologies for computation, and cyber security for data protection. They have emphasized the use of ML algorithms in irrigation systems, plant health monitoring, weed detection and removal, and crop growth and production enhancement.

In agricultural applications, Raut et al. [13] recommended adopting IoT for fertigation, irrigation, and soil monitoring. They evaluated the soil for macronutrients including potassium, phosphorus, and nitrogen and developed an autonomous irrigation system. The levels of these nutrients were high, medium, and low, according to the data. The proper nutrients were subsequently applied via an irrigation system, decreasing the need for human interaction and increasing irrigation system efficiency.

In 2023, Rosero et al. [14] conducted research on the collection and transmission of data to a mechanized D-type rolling robot in a smart greenhouse using an unsupervised algorithm (K-Means) on a wireless sensor network (WSN). For farmers, the Raspberry Pi acts as a GUI and central centre. According to the authors, farmers need to modify their farming practices to align with robot specifications for certain uses.

Son et al. [15] have used accelerometers and gyroscopes along with machine learning to classify farm workers' falls and everyday activities. A standard dataset, SisFall, was compared with biomechanical data from forty farm workers in this investigation. They looked at the efficiency of supervised machine learning classifiers such as ANN, SVM, and KNN algorithms. This study emphasizes the necessity for youthful, physically fit workers in agriculture because the number of farmers is declining, and the number of old people is rising. Next-generation farming powered by IoT is inevitable, according to the study.

Sundaresan et al. [16] highlighted smart farming on crop recommendation, fertilizer recommendation, and autonomous watering. The systems make use of an open-access cloud platform, K-means clustering, sensors, and a variety of machine learning models. The crop recommendation system monitors soil and makes recommendations for appropriate fertilizer and doses based on KNN and NPK sensors. In addition, a disease prediction system is included; the authors recommended putting comparable systems into place to produce bananas, apples, rice, maize, grapes, and rice.

These existing systems pave the way to develop a new IoT based greenhouse monitoring and controlling system to enhance crop productivity.

6.3 GREENHOUSE STRUCTURE AND DESIGN

Different types of greenhouses are designed to accommodate various climates, purposes, and budgets. Some common greenhouse types are listed below:

6.3.1 Conventional Greenhouse

These are standalone constructions composed of materials such as glass or polycarbonate panels. They usually have a peaked roof and provide high insulation.

6.3.2 Lean-To Greenhouse

The building's roof is covered with greenhouse roofing material, facing south, and can accommodate plant benches with widths ranging from 7 to 12 feet. The layout maximizes sunlight exposure, requires minimal expenditure, and minimizes the need for roof supports. Power, water, and heat are readily available, making it an economical and efficient solution.

6.3.3 Even-Span Greenhouse

The even-span structure is a full-size, where the pitch and width of the two roof slopes are equal. This design is utilized for a tiny greenhouse that is built on flat ground. It has one gable end attached to a dwelling. Two or three rows of plant benches are placed in it. These are more expensive than lean-to greenhouses, but they can accommodate more plants and have design flexibility.

6.3.4 Uneven-Span Greenhouse

This kind of greenhouse is constructed in a mountainous area. Since the roofing varies in width, the structure can adjust to the slopes on either side of the hill.

6.3.5 Sawtooth Greenhouse

This type of greenhouse permits natural ventilation. A particular natural ventilation flow channel is created in a sawtooth greenhouse.

6.3.6 Glass Greenhouse

This kind of greenhouse provides great light transmission and durability. They might be costly to create and maintain, but they provide a beautiful setting for growing plants.

6.3.7 Polycarbonate Greenhouse

Greenhouses composed of polycarbonate panels are long-lasting, lightweight, and provide excellent insulation. They are more economic than glass greenhouses while providing superior UV protection.

6.3.8 Aquaponic Greenhouse

These greenhouses use a closed loop system that combines hydroponics, or soilless plant growth, with aquaculture, or fish rearing. Plants receive nutrients from fish faeces and use those resources to filter and purify the water for fish.

6.3.9 Geodesic Dome Greenhouse

Geodesic domes have a high structural strength and are resource efficient. They provide good ventilation and can survive heavy snowfall and strong winds.

6.4 IOT TECHNOLOGY IN GREENHOUSE

The Internet of Things (IoT) is a network of physically interconnected items, including buildings and cars, which use software, sensors, and network connectivity to interact and gather data. Its principal objective is to make device interaction and communication simple, enabling information sharing and operations without the need for human intervention.

Applications of IoT in numerous fields include industries, including smart homes, industrial automation, transportation, agriculture, and healthcare monitoring. IoT can increase productivity, enhance decision-making processes, and create new opportunities for innovation in a variety of industries by allowing things to communicate and make intelligent decisions based on the data they gather. IoT applications across multiple industries are shown in Figure 6.1.

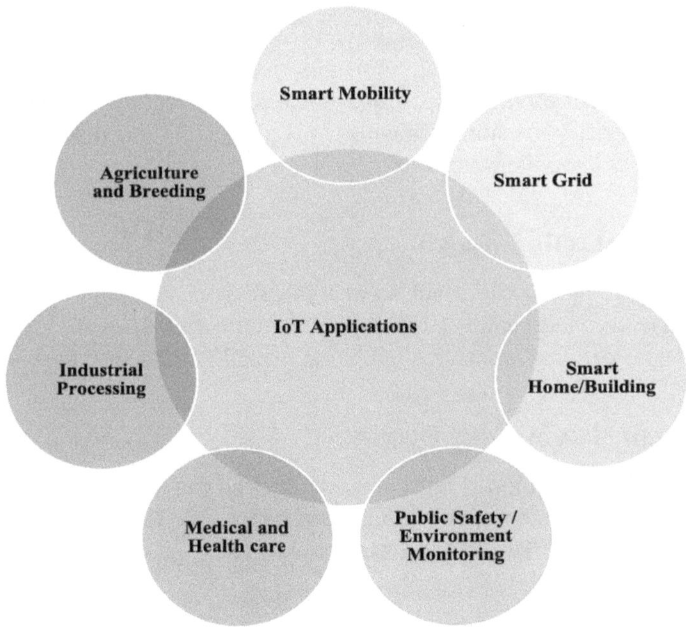

FIGURE 6.1 Applications of IoT.

Agriculture uses sensors to monitor dynamic data, agricultural indicators, and environmental conditions. The IoT provides various solutions in agricultural factors like soil, water, air, temperature, humidity, fertilizer, pest control, lighting management, and location monitoring. With the help of Arduino sensors and actuators, farmers can optimize irrigation techniques, boost crop output, and make data-driven decisions by gathering data on temperature, humidity, soil moisture, and other variables. The purpose of IoT and sensors in smart farming is depicted in Figure 6.2.

By combining data-driven methodologies and smart technologies, the IoT significantly influences and improves agriculture. Here are some key aspects of how IoT is used in agriculture:

6.4.1 Precision Farming

Sensor networks: IoT devices like drones, weather stations, and sensors can be used in the field to collect instantaneous information on crop health, weather patterns, and soil conditions.

Data Analysis: By analysing the gathered information, insights are gained regarding the optimal time to plant, when to water, and how much fertilizer and pesticide to use.

6.4.2 Smart Irrigation

- ***Soil Moisture Sensor:*** IoT-enabled soil moisture sensors can track soil moisture content and send information to irrigation systems. This makes it possible for irrigation to be done precisely and automatically, avoiding overwatering and saving water.
- ***Weather Forecast Integration:*** IoT devices can integrate weather forecasts in order to modify irrigation schedules in response to impending weather, which guarantees effective water use.

6.4.3 Health and Behaviour Tracking

Livestock can have IoT devices, such as wearable sensors, fitted to them to track their location, behaviour, and health. Farmers may better oversee breeding programmes, identify illnesses early and manage their herds overall with the use of this data.

6.4.4 Satellite Imaging and Drones

IoT-connected drones equipped with imaging devices can monitor large crop areas. With the use of this information, farmers are better able to evaluate the health of their crops, pinpoint problem regions, and decide on pest management and harvesting strategies.

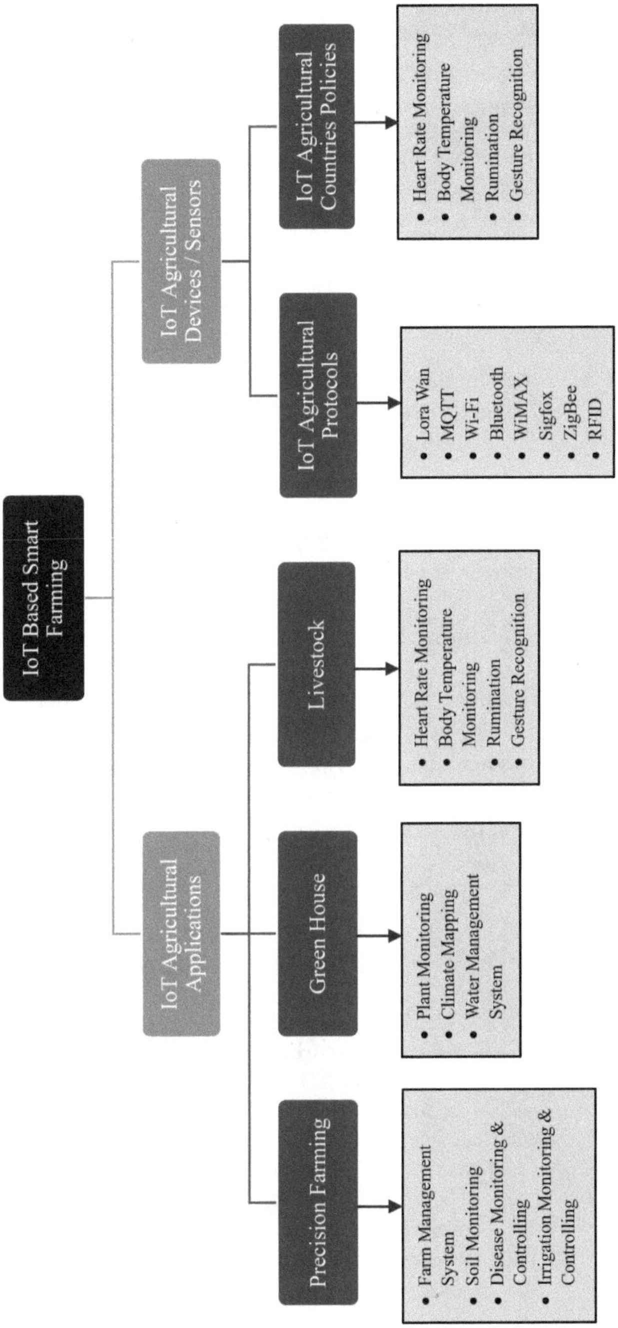

FIGURE 6.2 IoT in Smart Farming.

6.4.5 RFID and GPS Technology

This includes monitoring the transportation and storage conditions of crops, ensuring timely delivery to markets and reducing waste.

6.4.6 Climate Control

IoT devices in greenhouses can track and manage various environmental factors like temperature, humidity, and light. By facilitating the creation of ideal growing conditions for plants, automation contributes to increased agricultural yields and improved crop quality.

6.4.7 Telematics

IoT technology is integrated into agricultural machinery, allowing farmers to remotely monitor equipment performance, track usage, and schedule maintenance.

6.4.8 IoT Communication Protocol

The IoT requires a mandate communication channel through which the data can be transported, retrieved, stored, and analysed at various time. The standardized techniques used by IoT devices to interact with each other are known as IoT communication protocols. Modernizing and optimizing farming processes in agriculture are greatly aided by these guidelines. They make it easier for different smart equipment and sensors to seamlessly share data, which raises agricultural operations' overall production and efficiency. They also facilitate decision-making by sending data for analysis, enabling farmers to decide on crop management, pest control, and resource allocation with knowledge. Table 6.1 describes the various communication protocols in IoT for data transmission between the devices.

TABLE 6.1
IoT Communication Protocols

Protocol	Remarks
Low Range Wide Area Network Protocol (LoRaWAN)	Ensuring various operators' compatibility is the main goal of the LoRaWAN protocol. In order to increase crop productivity and reduce hazards, this framework has been developed.
Message Queue Telemetry Transport Protocol (MQTT)	Sensor data has been transmitted and received using MQTT. The irrigation issue has been fixed by employing the MQTT protocol to regulate the water pump's performance and provide information about soil moisture levels and pump status to the user's website and mobile application.
Radio-Frequency Identification (RFID)	It determines environmental factors including moisture content and temperature. An RFID tag is utilized to locate and track crop statistics.

TABLE 6.1 (Continued)
IoT Communication Protocols

Protocol	Remarks
sigFox	It is utilized for long-distance communications. Animal pastures have been localized for the entire summer using SigFox. Ranchers can trace the whereabouts of their cattle with the use of Sigfox.
ZigBee	It provides a long-range battery life. Due to high-speed data transfer, it is applicable in agriculture for rapid throughput.
WiFi	WiFi broadens the use of ad hoc networks to link different kinds of agricultural devices through heterogeneous architectures.
Bluetooth	A low-cost, low-power technique for communication over small distances, like 8 to 10 meters. It is applicable to various levels of agriculture.
Worldwide Interoperability for Microwave Access (WiMAX)	WiMAX ensures long range communication. Many agricultural applications can benefit from the usage of WiMAX, including water pump management, gate and light control, crop area border monitoring, and remote diagnostics of farming equipment.
5G Communication	The goals of 5G wireless technology are to provide more users with more consistent user experiences, super low latency, vast network capacity, faster multi-Gbps peak data rates, and better dependability.

6.5 MACHINE LEARNING IN A GREENHOUSE

The procedures listed below are frequently followed by farmers when doing agricultural activities.

- Crop selection
- Preparation of land
- Sowing of seeds
- Water irrigation & fertilization
- Maintenance of crop
- Crop harvesting
- Post harvesting activities

Various criteria must be examined while carrying out the above-mentioned procedure in order to increase crop output. The parameters to be considered are listed in the Table 6.2.

6.5.1 Machine Learning in Agriculture

Machine learning (ML) techniques are extremely beneficial for greenhouse farming to increase the crop yield, quality and resource efficiency. Figure 6.3 shows applications

TABLE 6.2
Parameters considered for Farming

Stage	Activities / Factors
Pre-harvesting	Quality of the soil and seeds, application of fertilizer and pesticides, selection of cultivars, genetic and environmental factors, irrigation and identification of weeds and diseases.
Harvesting	Crop size, skin colour, quality, crop detection and categorization.
Post harvesting	Crop-related factors include humidity, temperature, weed growth, use of pesticides in post-harvest and crop handling to maintain quality

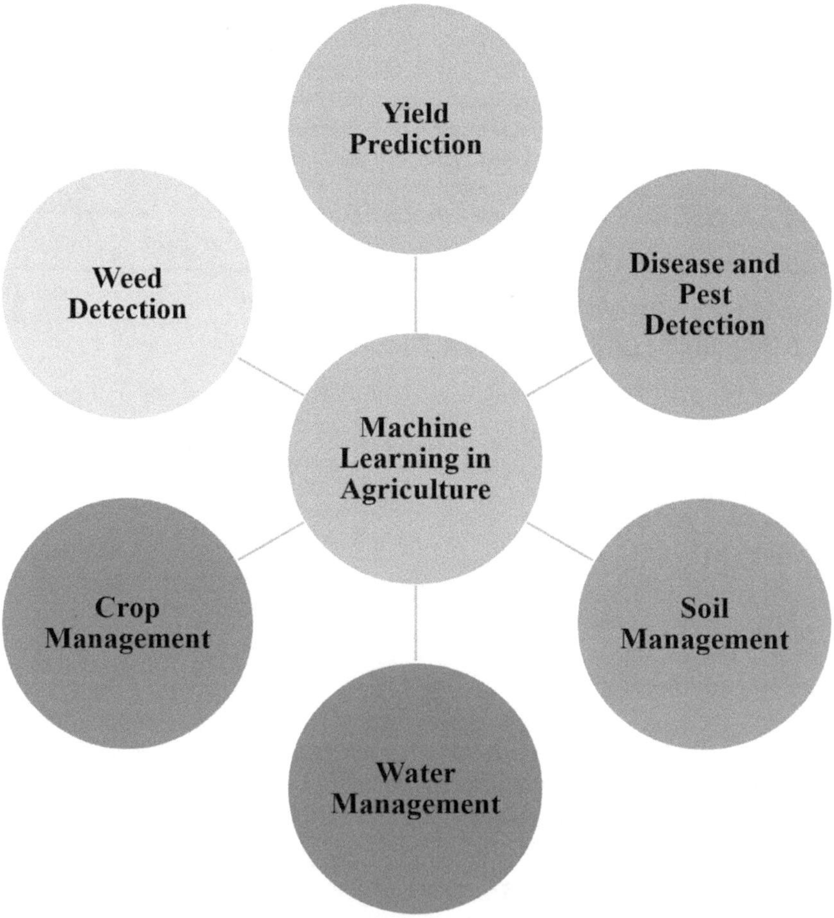

FIGURE 6.3 Machine Learning in Agriculture.

of ML in various aspects of agriculture. The various ways machine learning can be employed in greenhouse farming are:

Predictive Analytics: ML algorithms can predict the future greenhouse conditions by analysing past data on humidity, temperature, soil moisture, and light intensity. Such forecasts can assist farmers anticipate and manage future problems, such as plant diseases or poor growth conditions.

Crop Monitoring and Management: Data from sensors placed throughout the greenhouse may be interpreted by machine learning algorithms to track crop health in real time. ML algorithms can identify early indicators of illness, pests, or nutritional deficiencies by analysing this data, enabling farmers to promptly implement the necessary corrective measures.

Automated Control Systems: ML algorithms are capable of controlling many greenhouse environmental factors, including humidity, temperature and irrigation, by utilizing historical trends and real-time sensor data. Automated control systems can optimize growth conditions for individual crops, resulting in higher production and resource efficiency.

Optimized Resource Allocation: ML algorithms optimize the utilization of greenhouse resources like water, fertilizer, and energy to maximize crop quantity and quality while minimizing waste and lowering production costs.

Crop Variety Selection: ML models can analyse historical data on crop performance under different environmental conditions to recommend the most suitable crop varieties for specific greenhouse settings. This can help farmers select varieties that are well-adapted to local conditions and have higher yield potential.

6.5.3 Machine Learning Algorithms

Without the need for specialized programming, computers can identify patterns, make predictions, or develop decisions based on data by using computational models known as machine learning (ML) techniques. The classification of ML algorithms is described in Figure 6.4.

6.6 CLOUD TECHNOLOGY IN THE GREENHOUSE

Cloud technology is essential to modernize agriculture and to promote sustainability and environmental stewardship. In agriculture, cloud computing services and platforms are used to enhance various aspects of agricultural practices, management, and decision-making. It entails using distant servers housed online for the purpose of storing, handling, and processing agricultural applications and data. Cloud has the capability to store huge data securely, including weather patterns, soil conditions, crop yield data, and machinery performance metrics. Centralized data management is made possible by the fact that this data is accessible from any location with an internet connection. Advanced data analytics and processing can be done, which allow farmers to get valuable insights from their data. Based on historical and current

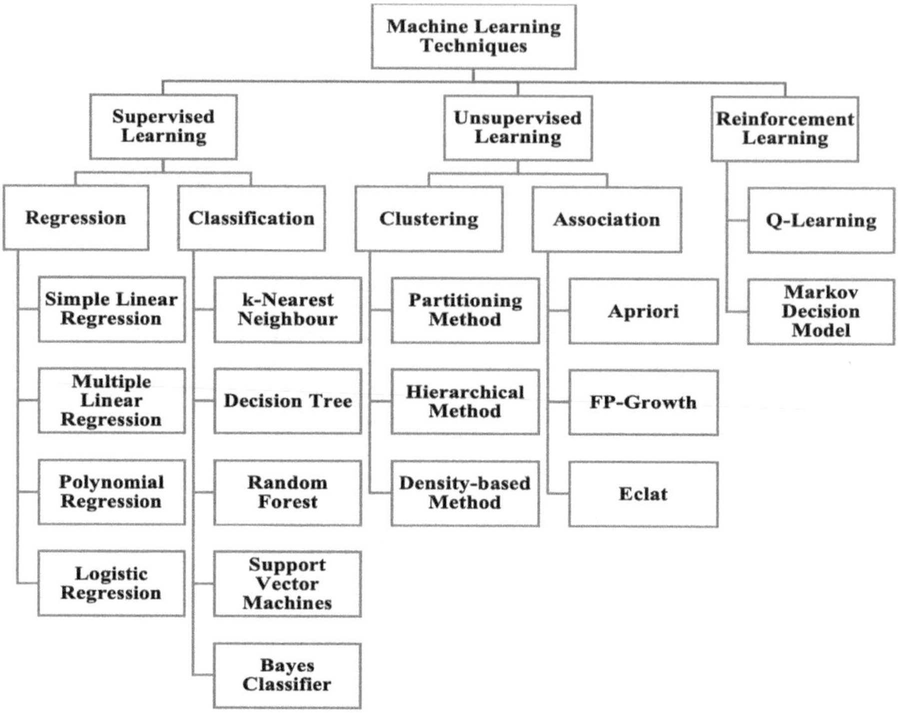

FIGURE 6.4 Classification of Machine Learning Algorithms.

data, this includes the best times to plant, disease outbreaks, insect infestations, and predictive analytics for crop yields.

6.7 MOBILE APPLICATION FOR THE GREENHOUSE

Farmers can access information through mobile applications. These apps assist farmers in making well-informed choices on pest control, irrigation, planting, and harvesting. Additionally, they provide planting guidelines, databases of crop kinds, and ideal growth conditions, all of which help businesses maximize yields and profit in a sector that is changing quickly.

Mobile applications offer pest and disease databases, symptom recognition guides, and integrated pest management (IPM) strategies for farmers. They provide recommendations for soil amendments, fertilization schedules, and nutrient management. These apps integrate data from sensors, drones, and images, allowing farmers to map fields, track crop health, and optimize inputs, leading to improved resource efficiency and cost savings.

6.8 PROPOSED METHOD

The system model of the automatic greenhouse monitoring and controlling system is depicted in Figure 6.5. It shows the list of components along with the input, output

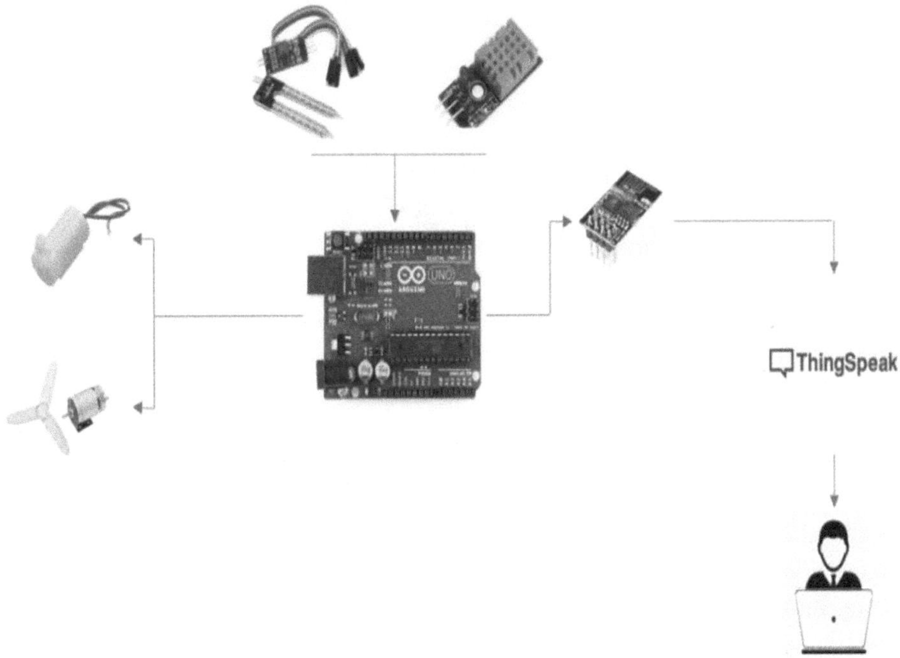

FIGURE 6.5 *IsAGHMCS* – System Design.

and the processing unit. The system uses sensors to gather data on a variety of environmental factors. If the actual values exceed the threshold values, the system triggers the actuators. Following that, the data is kept in a cloud database, making it accessible from anywhere at any time using a mobile application.

When the temperature goes beyond the set threshold, the fan activates automatically. If the soil's moisture level drops, the water pumping engine is automatically turned on. The automated fertilizer is activated if the soil's fertility declines. A smartphone application is being developed to obtain greenhouse environment reports.

6.8.1 IoT Technology in IsAGHMCS

The suggested automated greenhouse monitoring and controlling system for boosting agricultural output using IoT at a lower cost is schematically represented in Figure 6.6. Simultaneous monitoring of temperature, humidity, and soil moisture is done, and the data is recorded in the "ThingSpeak" cloud via a Wi-Fi module. To always keep an eye on the greenhouse conditions and from any location, an Android smartphone app has been developed.

Sensors used to monitor the environmental parameters in *IsAGHMCS* are listed in Table 6.3.

Figure 6.7 represents the circuit diagram for the automatic monitoring and controlling of a greenhouse environment. The connections are specified as follows:

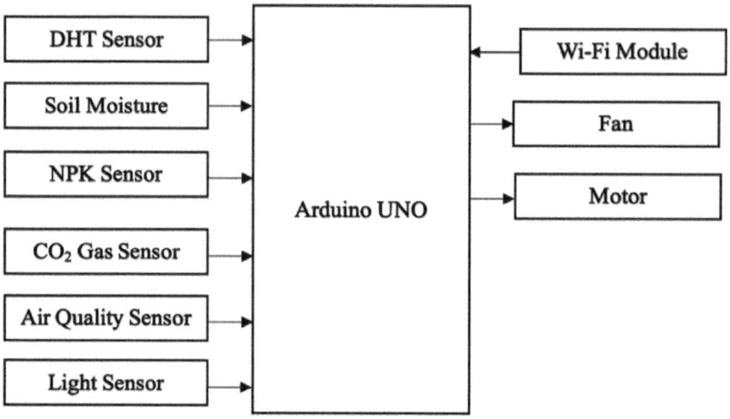

FIGURE 6.6 *In-Situ* Greenhouse Monitoring and Controlling System.

TABLE 6.3
IoT-based smart control systems in *IsAGHMCS*

Name of the Sensor	Operations
pH Sensor	Used to keep track of the precise amount of nutrients in the soil.
Gas Sensor	This sensor employs infrared light to precisely assess the level of harmful chemicals in cattle and greenhouses.
Motion Detector Sensor	It is used to track when an unwanted object is moving across a field or farm and sends alerts to farmers so they may take prompt action to prevent crop loss.
Ultraviolet Sensor	Monitors UV radiation to ensure that crops develop effectively.
Passive Infrared (PIR) Sensors	To monitor an individual's movement inside the field, a motion detector is placed. In addition, the sensor is capable of sensing light. It analyses crop development by converting rising temperature into voltage.
Soil Moisture Sensor	Monitors the amount of moisture and water present across the field.
Temperature Sensor	Temperature variations in the soil have an impact on how well moisture and nutrients are absorbed. This sensor is used to track humidity, CO_2, and temperature.
Humidity Sensor	Pollination, photosynthesis, and the growth of plant leaves are all negatively impacted by humidity. In order to determine the relative humidity, this sensor takes direct measurements of the temperature and moisture content of the air.
CO_2 Sensors	Used to track and identify the presence of carbon dioxide by absorbing infrared light at a certain wavelength.
NPK Sensor	Determines the amount of phosphorus (P), potassium (K), and nitrogen (N) in the soil.

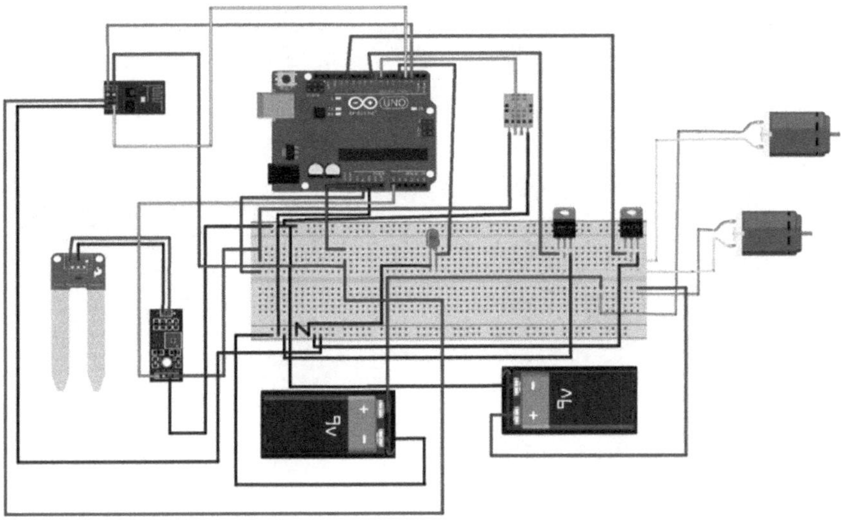

FIGURE 6.7 *IsAGHMCS* – Circuit Diagram.

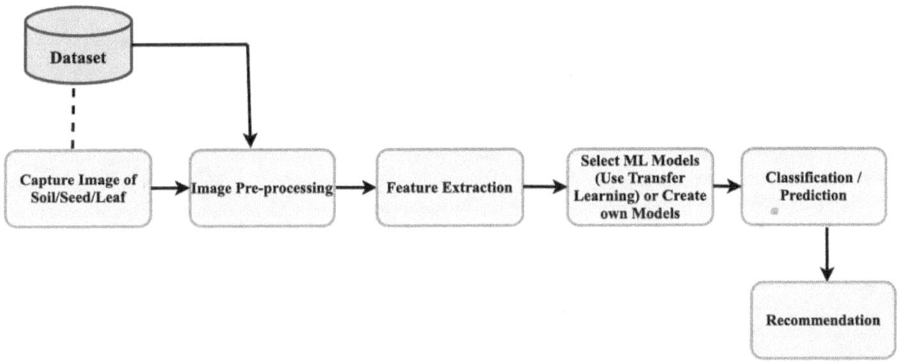

FIGURE 6.8 Machine Learning – *IsAGHMCS.*

6.8.2 Machine Learning in IsAGHMCS

The greenhouse farm should have the proper configuration and monitoring of the model parameters, such as temperature, humidity level, soil moisture content, and soil fertility. The data collected from the farm can be either numerical data or image. Once data collection is finished, pre-processing is carried out to offer an appropriate input for the ML models. For the image, most predominant features are extracted using various feature extraction techniques for the further processing. These features will be given as an input to any of the ML models. These ML models are more effective and produce accurate results which help the farmers for decision-making. Plant disease prediction, pest detection, weed detection, and soil fertility management are done using various ML algorithms. Figure 6.8 depicts the ML model for *IsAGHMCS.*

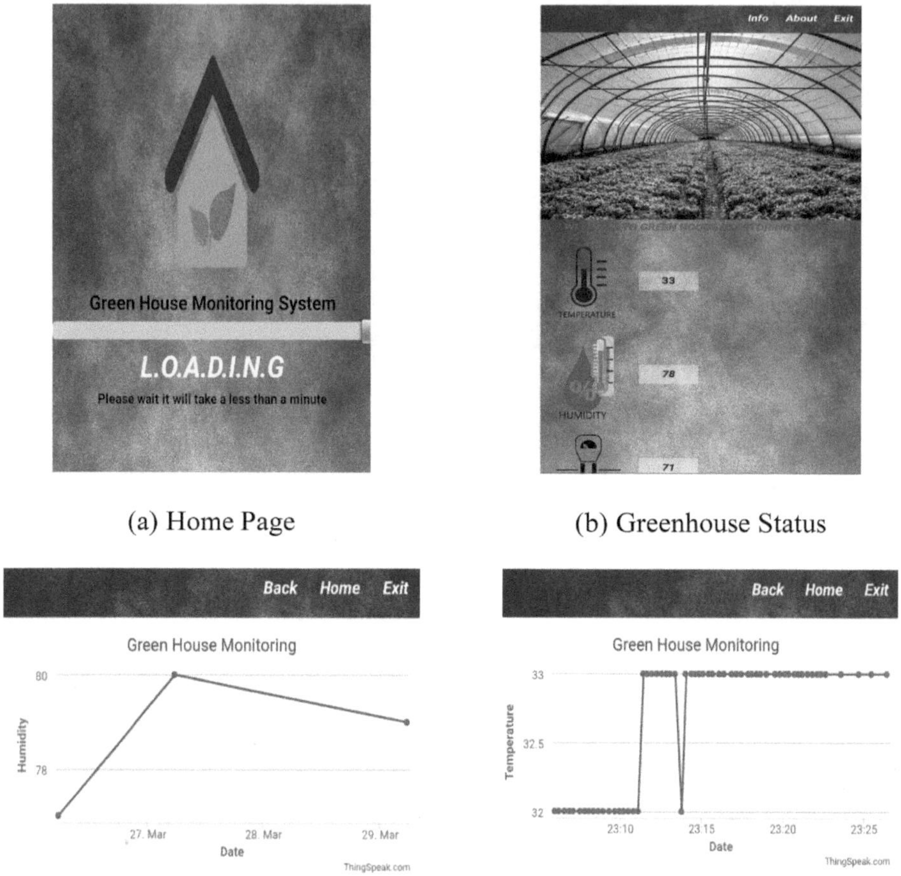

(a) Home Page (b) Greenhouse Status

(c) Humidity Level (d) Temperature Level

FIGURE 6.9 *IsAGHMCS* – VivasayI Mobile App.

Figure 6.9 shows the mobile application for monitoring and controlling the greenhouse environment.

6.9 CHALLENGES

IoT and ML have enormous benefits for the agricultural sector. However, there are a few obstacles to overcome when applying ML and IoT algorithms in the agricultural sector. These are as follows:

Data: To construct ML models, data is the most basic prerequisite. Many researchers find that gathering data is challenging for several reasons, such as a deficiency of data, data that is unavailable in the format required, bad quality data and data that can include extraneous factors. Also, data received from the sensors should be reliable.

Data pre-processing: Since there are many issues with data, it is necessary to use a variety of pre-processing methods to prepare the data for model training, testing, and validation. This might be a laborious procedure.

Selection of ML algorithms: There are many accessible ML algorithms, making it challenging to select the best algorithm for creating a customized ML model. Random selection is frequently necessary, or one might choose the most suited method by comparing the output of several algorithms. This kind of trial and error might cause the deployment of the model to take longer.

6.10 CONCLUSION

The fusion of IoT and ML in greenhouse management offers a transformative solution for agricultural efficiency and productivity. Through real-time data collection and analysis, it enables precise environmental adjustments tailored to plant needs, optimizing growth conditions, and resource utilization. This transformative technology empowers farmers with actionable insights, proactive management practices, and facilitating informed decision-making. The implementation of such innovative solutions promises enhanced crop yields, sustainability, and economic viability in modern agriculture. Ultimately, the implementation of *IsAGHMCS* improves agricultural productivity and advances food production sustainability.

REFERENCES

1. Qazi, S., Khawaja, B. A., & Farooq, Q. U. (2022). IoT-Equipped and AI-Enabled Next Generation Smart Agriculture: A Critical Review, Current Challenges and Future Trends. In IEEE Access (Vol. 10, pp. 21219–21235). Institute of Electrical and Electronics Engineers (IEEE). https://doi.org/10.1109/access.2022.3152544
2. AlZubi, A. A., & Galyna, K. (2023). Artificial Intelligence and Internet of Things for Sustainable Farming and Smart Agriculture. In IEEE Access (Vol. 11, pp. 78686–78692). Institute of Electrical and Electronics Engineers (IEEE). https://doi.org/10.1109/access.2023.3298215.
3. Farooq, M. S., Riaz, S., Abid, A., Abid, K., & Naeem, M. A. (2019). A Survey on the Role of IoT in Agriculture for the Implementation of Smart Farming. In IEEE Access (Vol. 7, pp. 156237–156271). Institute of Electrical and Electronics Engineers (IEEE). https://doi.org/10.1109/access.2019.2949703
4. Farooq, M. S., Javid, R., Riaz, S., & Atal, Z. (2022). IoT Based Smart Greenhouse Framework and Control Strategies for Sustainable Agriculture. In IEEE Access (Vol. 10, pp. 99394–99420). Institute of Electrical and Electronics Engineers (IEEE). https://doi.org/10.1109/access.2022.3204066
5. Farooq, M. S., Riaz, S., Helou, M. A., Khan, F. S., Abid, A., & Alvi, A. (2022). Internet of Things in Greenhouse Agriculture: A Survey on Enabling Technologies, Applications, and Protocols. In IEEE Access (Vol. 10, pp. 53374–53397). Institute of Electrical and Electronics Engineers (IEEE). https://doi.org/10.1109/access.2022.3166634

6. Lin, F., Weng, Y., Chen, H., & Zhuang, P. (2021). Intelligent Greenhouse System based on Remote Sensing Images and Machine Learning Promotes the Efficiency of Agricultural Economic Growth. In Environmental Technology & Innovation (Vol. 24, p. 101758). Elsevier BV. https://doi.org/10.1016/j.eti.2021.101758
7. Aburasain, R. Y. (2024). Enhanced Black Widow Optimization with Hybrid Deep Learning Enabled Intrusion Detection in Internet of Things-Based Smart Farming. In IEEE Access (Vol. 12, pp. 16621–16631). Institute of Electrical and Electronics Engineers (IEEE). https://doi.org/10.1109/access.2024.3359043
8. Astillo, P. V., Kim, J., Sharma, V., & You, I. (2020). SGF-MD: Behavior Rule Specification-Based Distributed Misbehavior Detection of Embedded IoT Devices in a Closed-Loop Smart Greenhouse Farming System. In IEEE Access (Vol. 8, pp. 196235–196252). Institute of Electrical and Electronics Engineers (IEEE). https://doi.org/10.1109/access.2020.3034096
9. Ayaz, M., Ammad-Uddin, M., Sharif, Z., Mansour, A., & Aggoune, E.-H. M. (2019). Internet-of-Things (IoT)-Based Smart Agriculture: Toward Making the Fields Talk. In IEEE Access (Vol. 7, pp. 129551–129583). Institute of Electrical and Electronics Engineers (IEEE). https://doi.org/10.1109/access.2019.2932609
10. Pak, J., Kim, J., Park, Y., & Son, H. I. (2022). Field Evaluation of Path-Planning Algorithms for Autonomous Mobile Robot in Smart Farms. In IEEE Access (Vol. 10, pp. 60253–60266). Institute of Electrical and Electronics Engineers (IEEE). https://doi.org/10.1109/access.2022.3181131
11. Huang, K., Shu, L., Li, K., Yang, F., Han, G., Wang, X., & Pearson, S. (2020). Photovoltaic Agricultural Internet of Things Towards Realizing the Next Generation of Smart Farming. In IEEE Access (Vol. 8, pp. 76300–76312). Institute of Electrical and Electronics Engineers (IEEE). https://doi.org/10.1109/access.2020.2988663
12. Ndjuluwa, L. N. P., Adebisi, J. A., & Dayoub, M. (2023). Internet of Things for Crop Farming: A Review of Technologies and Applications. In Commodities (Vol. 2, Issue 4, pp. 367–381). MDPI AG. https://doi.org/10.3390/commodities2040021
13. Raut, R., Varma, H., Mulla, C., & Pawar, V. R. (2017). Soil Monitoring, Fertigation, and Irrigation System Using IoT for Agricultural Application. In Intelligent Communication and Computational Technologies (pp. 67–73). Springer Singapore. https://doi.org/10.1007/978-981-10-5523-2_7
14. Rosero-Montalvo, P. D., Gordillo-Gordillo, C. A., & Hernandez, W. (2023). Smart Farming Robot for Detecting Environmental Conditions in a Greenhouse. In IEEE Access (Vol. 11, pp. 57843–57853). Institute of Electrical and Electronics Engineers (IEEE). https://doi.org/10.1109/access.2023.3283986
15. Son, H., Lim, J. W., Park, S., Park, B., Han, J., Kim, H. B., Lee, M. C., Jang, K.-J., Kim, G., & Chung, J. H. (2022). A Machine Learning Approach for the Classification of Falls and Activities of Daily Living in Agricultural Workers. In IEEE Access (Vol. 10, pp. 77418–77431). Institute of Electrical and Electronics Engineers (IEEE). https://doi.org/10.1109/access.2022.3190618
16. Sundaresan, S., Daniel Johnson, S., Mani Bharathy, V., Mohan Pavan Kumar, P., & Surendar, M. (2023). Machine Learning and IoT-based Smart Farming for Enhancing the Crop Yield. In Journal of Physics: Conference Series (Vol. 2466, Issue 1, p. 012028). IOP Publishing. https://doi.org/10.1088/1742-6596/2466/1/012028

7 Tiny Ambient IoT in Precision Agriculture

Revolutionizing Farming Practices Through Efficient Monitoring and Management

Shruthi Ravikumar, Sneha Iyer, Madhumathi Ramasamy, and Sathya S

7.1 INTRODUCTION

The agricultural sector is undergoing a significant transformation with the integration of modern technologies. Precision agriculture, driven by advancements in IoT and data analytics, is revolutionizing traditional farming practices.

7.1.1 The Rise of Precision Agriculture

Traditional farming practices have played a vital role in sustaining humanity for millennia. However, with a growing global population and a finite amount of arable land, these practices are facing significant challenges. Resource scarcity, particularly of water and fertile land, is a pressing concern. Additionally, traditional farming methods can contribute to environmental problems like soil erosion, water pollution from excessive fertilizer use, and greenhouse gas emissions. Traditional farming faces various challenges, namely climate change, growth in population, resource scarcity, and environmental degradation. Technology plays a crucial role in addressing these challenges by providing innovative solutions to improve efficiency, productivity, and sustainability in agriculture. From automated machinery to satellite imaging and data analytics, technological advancements have revolutionized farming practices worldwide.

Precision agriculture emerges as a transformative approach to address these challenges. It leverages technology to reduce usage of resources, enhance crop yield [1], and decrease environmental impact. At the core of precision farming lies the concept of data-driven decision making [2]. By collecting and analyzing real-time data on various farm parameters, farmers can make informed choices about resource allocation, crop management practices, and overall farm operations.

DOI: 10.1201/9781003487005-7

7.1.2 The Need for Real-Time Data

Effective decision making in precision agriculture hinges on the availability of real-time data. Traditional approaches to data collection, such as manual monitoring and periodic soil sampling, are frequently characterized by their labor-intensive nature, time-consuming processes, and limited ability to offer comprehensive insights. These methods cannot capture the dynamic nature of agricultural environments, where conditions can change rapidly.

For instance, relying solely on manual soil moisture checks might not reveal crucial fluctuations that could impact crop health. Similarly, infrequent sampling of nutrient levels in the soil could lead to missed opportunities to optimize fertilizer application. Real-time data collection allows for continuous monitoring and immediate response to changes in the agricultural environment.

7.1.3 Introducing Tiny Ambient IoT Devices

Tiny Ambient Internet of Things (IoT) devices represent a groundbreaking innovation in data collection for precision agriculture. These devices are compact, energy-efficient sensors equipped with wireless connectivity capabilities, and low-cost intelligent sensors strategically deployed across agricultural landscapes. Their small size and unobtrusive design make them ideal for seamless integration into existing farming practices without disrupting crops or soil. Unlike traditional bulky monitoring equipment, Tiny IoT devices can be easily deployed in large numbers, creating a dense network of data collection points. This allows for comprehensive monitoring of various farm parameters across the entire field, providing a more holistic view of the agricultural environment.

This chapter lays the groundwork for understanding the importance of precision agriculture and the limitations of traditional data collection methods. Tiny Ambient IoT devices are a revolutionary solution for real-time data acquisition in agriculture. The following sections will delve deeper into the functioning of IoT in agriculture, explore the specific characteristics of Tiny IoT devices, and showcase their diverse applications in various farm management practices.

7.2 THE ROLE OF IOT IN PRECISION AGRICULTURE

IoT serves as a bridge between traditional agricultural practices and modern technological advancements. By seamlessly integrating smart devices and sensors into farming operations, IoT enables precise monitoring and management.

7.2.1 Transforming Traditional Farming

IoT has revolutionized various industries by enabling the interconnection of devices and objects through sensors, actuators, and internet connectivity. In the agricultural sector, IoT is poised to usher in a new era of intelligent farming by leveraging a network of interconnected sensors and devices. IoT expedites the collection of on-time

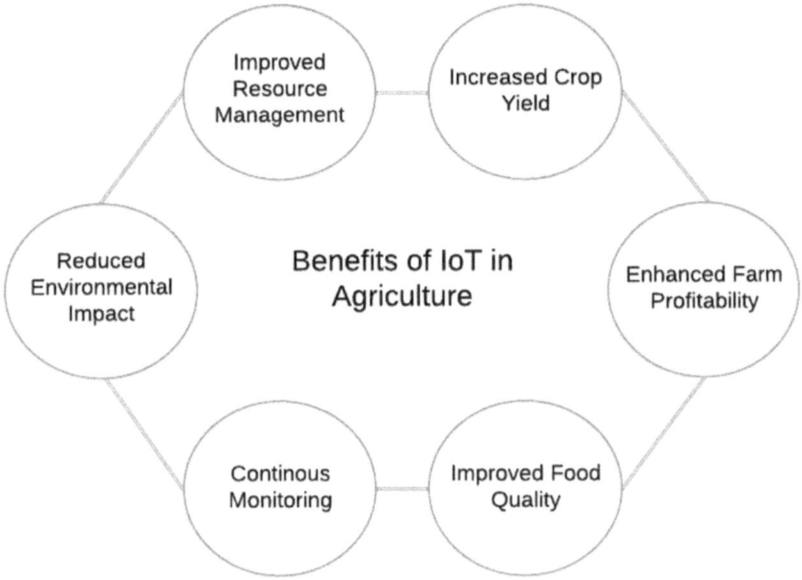

FIGURE 7.1 Agricultural IoT Benefits.

data in diverse farm parameters, including soil temperature, moisture, humidity, light levels, and even crop health indicators [3].

This instantaneous data equips farmers with the knowledge needed to make well-informed decisions across various aspects of farm management. Similarly, monitoring temperature and humidity can help in predicting and preventing the spread of crop diseases.

7.2.2 Benefits of IoT in Precision Agriculture

The integration of IoT technology into precision agriculture presents numerous advantages for farmers, the ecosystem, and consumers alike as shown in Figure 7.1. By leveraging IoT solutions, farmers can effectively manage resources by continuously monitoring real-time data related to water usage, fertilizer requirements, and pest activity. This enables them to precisely allocate resources, minimizing waste and maximizing efficiency in agricultural practices. Moreover, IoT facilitates the improvement of crop yield by enabling proactive measures to address possible concerns like nutrient deficiencies or pest infestations. Through the implementation of precision irrigation, targeted fertilizer application, and optimized operation of farm machinery, IoT aids in enhancing crop health and productivity.

7.3 TINY IOT DEVICES AND THEIR CHARACTERISTICS

In the swiftly expanding realm of the IoT, a notable trend is the rise of small-scale IoT devices. These compact gadgets are engineered to be energy-efficient, compact, and often feature-rich, catering to a broad spectrum of applications across various sectors. Let's delve into the defining traits of these diminutive devices and discover their characteristics.

- **Compact Design**: Small-scale IoT devices are characterized by their diminutive size. Leveraging advancements in miniaturization technologies like System-on-Chip (SoC) designs and Microelectromechanical Systems (MEMS), manufacturers can integrate significant functionality into compact packages. This compact design enables seamless integration into diverse environments, from wearable tech to industrial sensors.
- **Low Energy Consumption**: Efficient power management is critical for small-scale IoT devices, especially those installed in remote areas or operated by batteries. These devices typically utilize low-power components, optimized firmware, and power-saving modes to conserve energy. By minimizing energy consumption, they can operate for extended periods without frequent battery changes or external power sources.
- **Wireless Connectivity**: Connectivity is fundamental to IoT, and small-scale devices are no exception. They often incorporate wireless communication standards like Zigbee, Bluetooth Low Energy (BLE), or Wi-Fi, facilitating smooth data exchange with other devices or cloud platforms. These wireless capabilities enable real-time monitoring, remote control, and data-driven insights across diverse IoT ecosystems.
- **Sensor Integration**: Many small-scale IoT devices come equipped with an array of sensors tailored to specific applications. These sensors can measure parameters such as temperature, humidity, motion, light, and environmental pollutants. By integrating sensors directly into the device, it can collect rich data streams for analytics, automation, and decision-making purposes.
- **Edge Computing Capabilities**: IoT devices can incorporate edge computing capabilities and perform localized data processing, filtering, and analysis tasks, reducing latency and dependence on cloud infrastructure. Edge computing empowers these devices to make intelligent decisions autonomously or in conjunction with centralized systems.

7.3.1 Soil Moisture, Temperature, and Humidity Sensors

Monitoring the moisture content and humidity of the soil is shown in Figures 7.2 and 7.3, allowing farmers to irrigate their crops more efficiently. Soil temperature sensors monitor soil temperature, which influences seed germination, nutrient availability, and microbial activity in the soil. This helps to avoid underwatering or overwatering, which can both damage crops.

7.3.2 Nutrient Level and pH Sensors

These sensors measure the concentrations of vital nutrients such as nitrogen, phosphorus, and potassium in the soil, helping farmers determine the optimal fertilizer application rates based on crop needs. This reduces waste and ensures that crops get

FIGURE 7.2 Mini Hygrometer Thermometer.

FIGURE 7.3 Soil Moisture, Temperature and EC Sensor.

the nutrients they need to thrive. Soil pH sensors gauge the alkalinity or acidity of the soil, enabling farmers to regulate pH levels for optimal crop growth and nutrient uptake.

7.3.3 Leaf Sensors

Leaf wetness sensors as shown in Figure 7.4 detect the presence of moisture and temperature on leaves, which can analyze the leaf status such as watering, moisturizing, dew, and frozen and be a detector in the spread of plant diseases. By monitoring leaf wetness, farmers can take steps to prevent disease outbreaks.

7.3.4 Smart Greenhouse Sensors

These sensors monitor environmental conditions inside greenhouses, including temperature, humidity, light intensity, and CO2 levels as shown in Figure 7.5, optimizing growing conditions for crops.

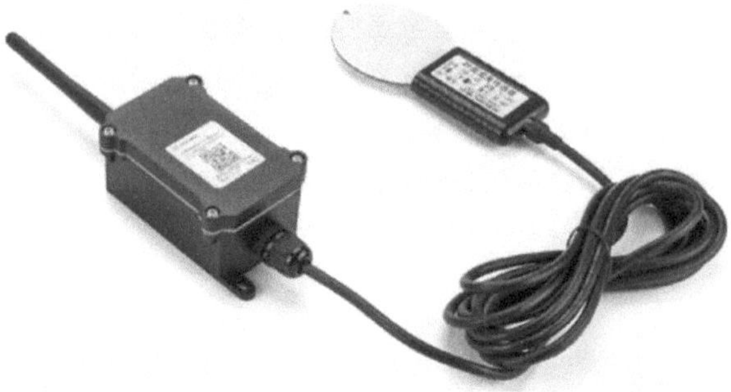

FIGURE 7.4 Leaf Sensor.

FIGURE 7.5 Carbon Dioxide Sensor.

7.3.5 Asset Tracking Sensor

These are used to monitor and maintain the equipment status in agriculture such as tractors, livestock, and other farm machinery from irrigation to harvest. The sensor shown in Figure 7.6 uses GPS enabled tracking system to track the equipment's identifies missing and damage equipment.

7.4 APPLICATIONS OF TINY IOT IN PRECISION AGRICULTURE

Tiny IoT devices, due to their small dimensions, low power consumption, and often low cost, find several applications in precision agriculture [4].

FIGURE 7.6 Mini GPS Tracker Asset Tracking Sensor.

7.4.1 Soil Monitoring

Tiny IoT sensors can be embedded in the soil to measure various parameters such as moisture content, pH levels, temperature, and nutrient levels. By providing precise data on soil conditions, Tiny IoT sensors empower farmers to optimize irrigation schedules, fertilizer usage, and overall soil health management. Tiny IoT sensors can provide farmers with precise data on soil nutrient levels, enabling them to tailor fertilizer applications to meet the specific requirements of the crop and avoid overuse or wastage of fertilizers. Precision agriculture allows for site-specific soil monitoring, where soil properties are assessed at a fine spatial scale within a field. This granular level of monitoring enables farmers to identify variability in soil characteristics and apply targeted management practices, such as variable rate application of inputs, to optimize crop performance and yield.

7.4.2 Crop Monitoring

Tiny IoT devices equipped with cameras or sensors can monitor crop growth and health in real-time. The sensors can identify subtle changes in plant health indicators, such as chlorophyll levels or leaf temperature variations, which may signal stress or nutrient deficiencies. They can detect factors like plant height, leaf color, and pest infestations. This data can assist farmers in making timely decisions regarding crop management, pest control, and harvest planning. IoT-enabled devices equipped with image recognition algorithms can identify signs of pest infestations or disease outbreaks. By analyzing images captured by cameras or sensors, farmers can detect pests, pathogens, or fungal infections early on, allowing for timely intervention strategies such as pest traps, biological control methods, or selective pesticide applications.

7.4.3 Weather Monitoring

Tiny IoT weather stations measure rainfall intensity, duration, and distribution patterns, providing farmers with essential information for irrigation scheduling, soil moisture management, and flood risk assessment. By tracking precipitation levels in real-time, farmers can optimize water usage, prevent waterlogging or drought stress, improve overall crop resilience and protect crops from adverse weather events. IoT-enabled devices monitor air and soil temperatures, helping farmers anticipate frost events, heatwaves, or temperature fluctuations that may impact crop growth and development. By implementing protective measures such as frost alarms, row covers, or irrigation systems, farmers can minimize frost damage, optimize growing conditions, and safeguard crop yields. Tiny IoT weather stations track wind speed and direction, which influence pollination, pesticide drift, and crop lodging risks. By monitoring wind conditions, farmers can adjust spraying operations, deploy windbreaks, or implement crop management practices to mitigate wind-related damage and optimize crop health. IoT sensors measure solar radiation levels, which affect photosynthesis, evapotranspiration, and crop growth rates. By monitoring solar radiation data, farmers can optimize planting times, crop spacing, and shading techniques to maximize sunlight exposure, enhance crop productivity, and improve overall yield and quality.

7.4.4 Livestock Monitoring

For livestock farming, Tiny IoT devices can be attached to animals to track their health, location, and behavior. These devices can monitor parameters like body temperature, activity levels, and feeding patterns. By analyzing this data, farmers can detect illnesses early, optimize feeding regimes, and improve overall animal welfare.

7.4.5 Precision Irrigation and Crop Spraying

Tiny IoT-enabled irrigation systems can deliver water precisely to crops based on real-time data from soil moisture sensors and weather forecasts. This guarantees that plants receive the appropriate amount of water when needed, reducing water wastage and maximizing crop yield. Also, tiny IoT devices mounted on drones or autonomous vehicles can spray pesticides or fertilizers precisely, targeting specific areas of the field based on crop health data collected from sensors. Tiny IoT devices can detect early signs of plant diseases and pest infestations through data analysis of environmental conditions and crop health indicators. This early detection enables farmers to enact precise interventions, like targeted spraying or biological control methods, to mitigate crop damage and yield losses.

7.4.6 Supply Chain Management and Optimization

IoT devices can track the movement of agricultural products from farm to market, monitoring factors like temperature, humidity, and handling conditions. This helps ensure product quality and safety throughout the supply chain, reducing spoilage and losses. By analyzing data collected from Tiny IoT devices throughout the agricultural supply chain, stakeholders can gain insights into market trends, demand patterns, and supply chain dynamics. This information helps optimize production schedules, distribution routes, and inventory management, ensuring timely delivery of agricultural products to consumers.

7.4.7 Remote Monitoring and Data-Driven Decision Making

Tiny IoT devices enable farmers to remotely monitor and control various agricultural operations, such as irrigation systems, greenhouse environments, and livestock facilities. Through mobile applications or web interfaces, farmers can adjust settings, receive alerts, and track performance in real-time, even when they're not physically present on the farm. By collecting and analyzing data from Tiny IoT sensors, farmers can enhance their decision-making process regarding crop selection, planting times, and harvesting strategies.

In conclusion, small-scale IoT devices represent a significant advancement in connected technologies. With their compact size, low energy consumption, and versatile capabilities, they are poised to drive innovation across industries, shaping the future of a truly interconnected world.

7.5 VARIOUS METHODS IN UTILIZATION OF IOT IN PRECISION AGRICULTURE

IoT has transformed precision agriculture by enabling real-time monitoring, analysis, and control of farming operations. This innovation enhances productivity, resource efficiency, and sustainability through advanced IoT devices and connected systems.

7.5.1 Assessment of Factors Influencing the Accuracy of High-Resolution Soil Property Mapping

In the realm of soil science, the creation of high-resolution maps for key properties such as soil organic carbon (SOC), clay content, and pH is crucial for various agricultural and environmental applications.

7.5.1.1 Training Sample Size Evaluation

The study embarked on an investigation varying the training sample size from 5 to 100 observations to discern its influence on map accuracy. Results demonstrated a diminishing return pattern, with the relative improvement in prediction accuracy diminishing exponentially with larger sample sizes. Notably, the most favorable median root-mean-square error (RMSE) values were attained with a sample size of

100, while even with a modest sample size of 10, models of acceptable quality (CCC > 0.65) were achieved.

7.5.1.2 Sampling Design Comparison

In order to assess the impact of different sampling designs, three distinct methodologies were scrutinized: simple random sampling (SRS), K-means sampling (KM), and conditioned Latin hypercube sampling (cLHS). The outcomes revealed a substantial superiority of cLHS and KM over SRS in terms of prediction accuracy, indicating the importance of a thoughtful sampling strategy in enhancing map precision.

7.5.1.3 Evaluation of Prediction Models

While increasing training sample size generally improves predictive performance, the selection of an appropriate sampling design and prediction model is equally crucial. These findings provide valuable insights for optimizing the generation of precise soil property maps, thereby contributing to informed decision making in agricultural and environmental management contexts [5].

7.5.2 Enhancing Precision Agriculture through IoT-Enabled Remote Monitoring and Control of Tractor Operations

An advanced system aimed at optimizing precision agriculture practices by harnessing the capabilities of IoT technology. By leveraging cloud computing and Wi-Fi-based networks, the system facilitates seamless communication with farmers in rural areas, enabling efficient management of tractor operations from remote locations. The research emphasizes the development of an IoT-based network architecture designed to enable two primary functionalities: remote control of tractor engine operations and real-time tracking and monitoring of engine status through a dedicated web interface. Central to the system is the Raspberry Pi-3, serving as the primary processing unit within the On-Board Unit OBU, responsible for aggregating and analyzing both engine status and positional data. The OBU is equipped with an Engine Control Sub Unit (ECSU) and a GPS tracking unit. Processed information is then relayed to the web server, where it is stored in a database for subsequent retrieval. [6].

7.5.3 Scalability and Performance Evaluation of (WSNs) Sensor Networks in Precision Agriculture

The transformative potential of the IoT in agriculture, particularly focusing on real-time monitoring of crop and livestock conditions. With an emphasis on enhancing precision agriculture practices, the research aims to assess the scalability and performance of WSNs in two distinct farming scenarios: olive tree farms and horse stables for training purposes. The study utilizes COOJA, a realistic WSN simulator, to model and simulate the performance of 6 LowPAN networks and the Routing protocol for low-power and lossy networks (RPL) in the agricultural scenarios identified. Simulation parameters are meticulously set for both stationary and mobile nodes, with particular attention given to examining the influence of node mobility on network performance. Key performance indicators such as radio duty cycle, average current consumption, and sensor network connectivity are analyzed to characterize the performance requirements in the agricultural settings under study. [7].

7.5.4 Leveraging Internet of Things for Precision Agriculture: A Paradigm Shift in Resource Management

The integration of IoT technology in precision agriculture revolutionizes resource management by enabling real-time monitoring and control of agricultural parameters. Sensor nodes deployed in the field collect data, which can be transmitted to the cloud for centralized monitoring and management. This empowers farmers and agricultural stakeholders to remotely access real-time data, facilitating timely decision making to optimize agricultural operations. In essence, the convergence of IoT and precision agriculture represents a paradigm shift in resource management practices. By employing the power of IoT, farmers can make data-driven decisions to enhance efficiency, productivity, and sustainability in agricultural practices [8].

7.5.5 Progress in IoT-enabled Precision Agriculture – Integrating Sensor Technologies, Actuators, and Operational Strategies

In modern agricultural practices, the synergistic integration of sensor technologies and actuators plays a pivotal role in optimizing resource allocation and enhancing operational efficiencies within agroecosystems. Sensors serve as the primary data capture mechanism, gathering information on various physical and environmental parameters such as meteorological conditions, crop growth stages, and soil characteristics. Meanwhile, actuators respond to sensor feedback, enabling proactive control over agricultural processes. The devices, capable of detecting and converting physical phenomena into actionable signals, represent a convergence of natural and engineered systems. In the deployment of sensor networks, careful consideration is given to selecting optimal field locations and placing sensors at depths corresponding to the active root zones of target crops. Strategic sensor placement provides valuable insights into soil moisture dynamics, enabling tailored irrigation management to meet crop needs and environmental conditions. The advent of IoT-enabled precision irrigation represents a paradigm shift in agricultural practices, offering autonomous, data-driven solutions for managing agricultural landscapes [9].

7.5.6 Enhancing Agricultural Efficiency and Sustainability through IoT Innovations

Management of pesticides and fertilizers, monitoring plant health, preventing diseases, monitoring irrigated agriculture, conserving soil, and tracing distribution networks are the key applications of IoT solutions in contemporary agriculture. These solutions collect environmental data from various contexts such as arable fields, orchards, and greenhouses, aiding in the evaluation of crop vigor and the regulation of environmental conditions. Aerial imagery assesses nitrogen content in large plantations, while autonomous robots optimize pesticide distribution in greenhouse environments. Industrial IoT platforms analyze parameters such as soil water content, nutrient levels, and weather conditions, facilitating informed decision making in agricultural practices. The integration of SBCs and Unmanned Aerial Vehicles (UAVs) in various agricultural applications is showcased, ranging from disease prevention to crop monitoring and irrigation control. UAVs equipped with multispectral sensors and cameras enable rapid and cost-effective crop surveillance, supporting tasks such as pest detection and fertilization optimization. [10]

7.5.7 Synergizing Precision Agriculture: Integrating Genetic Algorithms and IoT for Agriculture 4.0

A transformative approach to modern agricultural practices is rooted in the fusion of advanced technologies and traditional agricultural methods; this system represents a comprehensive strategy to enhance agricultural productivity, sustainability, and efficiency. At its core, Agriculture 4.0 symbolizes the integration of state-of-the-art technologies such as IoT, artificial intelligence (AI), and data analytics into agricultural processes. Complementing the IoT infrastructure, Genetic Algorithm (GA) methodologies provide a robust framework for optimizing complex agricultural systems. Inspired by the principles of natural selection and evolutionary biology, GA algorithms iteratively search for optimal solutions to agronomic challenges. By deploying sensor networks and data analytics frameworks, our system enables stakeholders to monitor environmental conditions, optimize resource usage, and implement adaptive management strategies. [11].

7.6 SYSTEM FOR TINY IOT USING WIRELESS SENSOR NETWORKS

To effectively monitor and manage soil conditions, the integration of WSNs plays a pivotal role in collecting real-time data from the agricultural field.

7.6.1 Wireless Sensor Networks

WSNs comprise a network of sensor nodes distributed throughout the agricultural field [12]. These nodes are equipped with various sensors capable of measuring soil parameters such as fertility, moisture content, pH level, and temperature. The sensors continuously collect data from the soil and wirelessly transmit it to a central gateway or base station.

7.6.2 Sensor Nodes

Each sensor node within the WSN is equipped with multiple sensors tailored to measure specific soil parameters. For soil fertility monitoring, sensors may include nutrient analyzers capable of detecting levels of fundamental elements such as nitrogen, phosphorus, and potassium. Additionally, other sensors such as soil moisture sensors, pH sensors, and temperature sensors provide complementary data for comprehensive soil analysis.

7.6.3 Data Transmission

The sensor nodes employ wireless communication protocols to transmit the collected data to a central gateway or base station. These protocols may include Zigbee, LoRa WAN, or Wi-Fi, depending on factors such as range, data rate, and power consumption requirements. The choice of protocol ensures efficient and reliable data transmission over long distances within the agricultural field.

7.6.4 Cloud-Based Platform

The transmitted data is related to a cloud-based platform for storage, processing, and analysis. Cloud computing infrastructure provides scalability and accessibility, allowing stakeholders to access real-time soil data from anywhere with an internet connection. Cloud-based platforms offer advanced data analytics capabilities for deriving actionable insights from the collected data.

7.6.5 Mobile Application

A mobile application serves as the interface for farmers and stakeholders to access and visualize the soil data collected by the system. Through the application, users can view real-time soil parameters, receive notifications and alerts, and access historical data for trend analysis. The application may also include features such as predictive modeling and decision support systems to assist farmers in making informed decisions about soil management practices.

7.6.6 IoT-Based Software System

The core of the proposed system is an IoT-based software platform that orchestrates the collection, transmission, storage, and analysis of soil data. The software employs algorithms and machine learning techniques to process the raw sensor data, identify patterns, and generate insights. Additionally, the software incorporates decision-making logic to recommend optimal fertilizer quantities based on soil nutrient levels and crop requirements.

7.7 TINY IOT CHALLENGES AND CONSIDERATIONS

Implementing Tiny IoT devices in agriculture comes with several challenges and factors that must be considered to guarantee successful deployment and operations [13].

7.7.1 Power Supply and Energy Efficiency

Tiny IoT devices often operate in remote agricultural environments where access to power sources may be limited. Ensuring a reliable and sustainable power supply for these devices is a significant challenge. Battery-powered solutions may require frequent maintenance and replacement, while solar or kinetic energy harvesting solutions need to be robust enough to withstand harsh outdoor conditions. Moreover, optimizing the energy efficiency of Tiny IoT devices is essential to prolong battery life and minimize power consumption, especially in low-power and intermittent connectivity scenarios.

7.7.2 Data Management and Connectivity

Collecting, transmitting, and managing data from Tiny IoT devices pose challenges related to connectivity, bandwidth, and data storage [14]. Agricultural fields may have limited or intermittent network coverage, making it challenging to establish reliable communication with remote sensors. Implementing efficient data transmission protocols, such as Lora WAN or MQTT, and utilizing edge computing techniques to process data locally can help overcome connectivity constraints. Furthermore, ensuring data privacy, security, and compliance with regulations such as GDPR is crucial when handling sensitive agricultural data.

7.7.3 Sensor Calibration and Reliability

Tiny IoT sensors used in agriculture must be accurate, reliable, and calibrated to provide meaningful data for decision making. Maintaining sensor accuracy and reliability in dynamic and harsh agricultural environments, characterized by temperature fluctuations, soil moisture variations, and exposure to dust and debris, is a significant challenge. Regular sensor calibration, quality assurance measures, and proactive maintenance practices are essential to ensure the reliability and integrity of data collected by Tiny IoT devices.

7.7.4 Integration and Interoperability

Integrating Tiny IoT devices into existing agricultural workflows, equipment, and management systems requires careful planning and consideration of interoperability standards and protocols [15]. Ensuring seamless communication and data exchange between different devices, platforms, and stakeholders is essential for optimizing farm operations and leveraging the full potential of IoT technology. Adopting open standards, APIs, and modular architectures can facilitate interoperability and enable flexible integration of Tiny IoT solutions with existing agricultural infrastructure and software ecosystems.

7.7.5 Environmental Durability

Agricultural environments can be harsh, with exposure to extreme temperatures, moisture, dust, and mechanical stress. Ensuring the durability and resilience of Tiny IoT devices to withstand these environmental factors is crucial for their long-term operation and reliability. Robust enclosure designs, ruggedized materials, and IP-rated protection against water and dust ingress are essential considerations for ensuring the longevity of IoT deployments in agriculture.

7.7.6 Scalability and Deployment Complexity

Scaling up IoT deployments across large agricultural operations or multiple farms can be complex and challenging. Managing many Tiny IoT devices, coordinating deployment logistics, and maintaining infrastructure across diverse geographical locations require careful planning and resource allocation. Standardizing deployment processes, leveraging remote management tools, and adopting scalable architectures can help streamline deployment and management at scale.

7.7.7 Cost and Return on Investment (ROI)

The initial investment and ongoing operational costs associated with deploying Tiny IoT devices in agriculture can be significant factors for consideration. Farmers must evaluate the cost-effectiveness and potential return on investment (ROI) of IoT solutions in terms of increased productivity, resource savings, and operational efficiencies. Balancing the upfront costs of hardware, installation, and infrastructure with the long-term benefits and value proposition of IoT-enabled precision agriculture is essential for adoption and sustainability.

7.8 CONCLUSION

In conclusion, the amalgamation of WSNs and Tiny IoT technology offers a groundbreaking solution for soil fertility management in agriculture. Through the integration of sophisticated sensor technology, wireless communication protocols, cloud-based platforms, and mobile applications, farmers can access real-time soil data, enabling precise decision making and the implementation of advanced agricultural practices. This innovative approach not only enhances crop yields but also fosters sustainable soil management practices, contributing to the long-term viability of agricultural ecosystems. As technology evolves, ongoing refinement and optimization of these systems are essential to meet the evolving needs and challenges of modern agriculture, ensuring food security and environmental resilience for future generations.

REFERENCES

1. Shadi Atalla, Saed Tarapiah, Amjad Gawanmeh, Mohammad Daradkeh, Husameldin Mukhtar, Yassine Himeur, Wathiq Mansoor, Kamarul Faizal Bin Hashim, Motaz Daadoo. 2023. IoT-Enabled Precision Agriculture: Developing an Ecosystem for Optimized Crop Management. Information. 14(4): 205.
2. Ravesa Akhter, Shabir Ahmad Sofi. 2022. Precision Agriculture using IoT Data Analytics and Machine Learning. Journal of King Saud University – Computer and Information Sciences 34(8): 5602–5618.
3. Ramasamy Madhumathi, Thangaraj Arumuganathan and Ravikumar Shruthi. 2021. A Survey on Wireless Sensor Networks and Instrumentation Techniques for Smart Agriculture. Mobile Computing and Sustainable Informatics. Springer Lecture Notes on Data Engineering and Communications Technologies. 68: 453–467.

4. Abdennabi Morchid, Rachid El Alami, Aeshah A. Raezah and Yassine Sabbar. 2024. Applications of Internet of Things (IoT) and Sensors Technology to Increase Food Security and Agricultural Sustainability: Benefits and Challenges. Ain Shams Engineering Journal 15(3): 102509.
5. Jonas Schmidinger, Ingmar Schröter, et al. 2024. Effect of training sample size, sampling design and prediction model on soil mapping with proximal sensing data for precision liming. Precision Agriculture 25(3). Berlin: Springer.
6. Prateek Shrivastava, Virendra Kumar Tewari, et al. 2023. IoT and Radio Telemetry based Wireless Engine Control and Realtime Position Tracking System for an Agricultural Tractor. International Journal of Discovering Internet of things 3(1). Publication no : 6. L 43926-023-00035-4.
7. Atalla Shadi, Tarapiah Saed, et al. 2023. IoT-Enabled Precision Agriculture: Developing an Ecosystem for Optimized Crop Management. Multidisciplinary Digital Publishing Institute, Vol. 14. Article: 205. doi:10.3390.
8. Sethuraman Muthukumarasamy, Chitraputran. Balakrishnan, et al. 2022. A Sensor-Based IoT for Precision Agriculture. International Journal of Creative Research Thoughts. ISSN: 2020-2882. 10(6): f15–f20.
9. S. Vinod Kumar, Chandra Deep Singh and Konga Upendar. 2022. IoT Based Precision Irrigation System in Agriculture. Central Journal of Applied Science and Technology. ISSN: 2231-0843
10. N. S. Abu, Wan Mohd Bukhari, C. H. Ong. 2022. Internet of Things Application in Precision Agriculture. International Journal of Robotics and Control. ISSN: 2715-5072. 3(3): 338–347.
11. Sayan Kumar Roy, and Debashis De. 2020. Genetic Algorithm based Internet of Precision Agricultural Things (IoT) for Agriculture 4.0. IEEE Internet of Things Journal. 18: 100201.
12. Antonis Tzounis, Nikolaos Katsoulas, Thomas Bartzanas, Constantinos Kittas. 2017. Internet of Things in agriculture, recent advances and future challenges. Biosystems Engineering 164: 31–48.
13. Porkodi Rajendran and Velumani Bhuvaneswari. 2014. The Internet of Things (IoT) Applications and Communication Enabling Techonology Standards: An Overview. International Conference on Intelligent Computing Applications. 324–329.
14. Zainab H. Ali, Hesham A. Ali, Mahmoud M. Badawy. 2015. Internet of Things (IoT): Definitions, Challenges and Recent Research Directions. International Journal of Computer Applications 128(1): 37–47.
15. Abdel Rahman H. Hussein. 2019. Internet of Things (IOT): Research Challenges and Future Applications. International Journal of Advanced Computer Science and Applications 10: 6.

8 Remote Sensing Applications in Precision Agriculture

Balaji Kannan, Janani N, Nirmala Devi M, and Ramanan M

8.1 INTRODUCTION

Food and fiber are two fundamental necessities of human being, which are met by agriculture [1]. Agriculture has been able to keep up with the rising demand for food and other agricultural goods because of the development of new farming techniques throughout the previous century, such as those used during the Green Revolution. Natural resources will undoubtedly be further strained as a result of rising food demand, population growth, and income levels. As the negative effects of agriculture on the environment come to light more frequently, new methods and strategies should be able to meet future food demands while preserving or minimizing agriculture's environmental footprint [2].

Precision farming, also known as site-specific management, refers to the practice of applying agronomic inputs across a farm, mainly fertilizers and other chemicals, at variable rates based on soil nutrients or chemical tests, soil textural changes, weed pressures and yield maps for each field in the farm [3]. Precision farming also has the potential for improving water use efficiency on large fields provided there is a quantitative understanding of what factors and where in the field they affect crop–water use. In most fields (e.g. > 40 ha) crop yields are noticeably variable [4]. The sources of this variation are related to the physical and chemical properties of the soil, pests, microclimate, genetic, and phonological responses of the crop and their interactions. The use of geospatial technologies has become part and parcel of precision agriculture.

The practice of precision agriculture (PA) involves the use of a variety of such remote sensing technologies to optimize agricultural inputs in order to boost agricultural output and lower input losses [5]. Throughout the past few decades, there has been a sharp expansion in the use of remote sensing technology for precision agriculture. The use of remote sensing has been encouraged in many PA applications, including crop monitoring, irrigation management, nutrient application, disease and pest management, and yield prediction, due to the previously sought availability of high resolution (spatial, spectral, and temporal) satellite images [6].

The tools and technologies offered by precision agriculture enable the identification of in-field soil and crop variability, providing a way to enhance sub-field level

DOI: 10.1201/9781003487005-8

farming techniques and optimize agronomic inputs [7]. With the help of variable-rate technology (VRT), soil and crop inputs can be applied at different rates depending on the site [8]. Today, field level data may be collected more easily and affordably with the use of sensing technologies that are both remote and ground-based. Maps of crop growth, crop diseases, weeds, crop nutrient deficits, and other crop and soil factors are necessary for the operational success of VRT [9]. As a result, remote sensing maps displaying crop and soil variability obtained by sensors mounted on satellites, aircraft or ground-based equipment have become a crucial component of VRT.

With remote sensing, crop yield can be mapped as well as soil parameters, crop species classification, crop water stress detection, weed and disease monitoring, and crop species classification [10]. The platforms (satellite, air, or ground) used for data collection; the number and width of spectral bands captured by the sensor (multi versus hyperspectral); and the spatial (high, medium, and low), temporal (hourly, daily, and weekly), and radiometric (8-, 12-, and 16-bit) resolutions at which sensors collect data all have an impact on the use of remote sensing in Precision Agriculture [9].

Precision farming has benefited from remote sensing due to the availability of high-resolution satellite images, including spatial, spectral, and temporal ones. Precision farming using remote sensing offers the tools and technologies for analysing data required for better farming methods and higher crop yield. Remote sensors installed on satellites, aircraft, unmanned aerial vehicles (UAV), or ground-based devices collect data.

Therefore, the advancement in geospatial technology tools like Remote Sensing, Geographic Information System, Global Positioning System, and Decision Support Systems have paved the way for better and more effective management of crop production at farm level.

8.2 REMOTE SENSING (RS) APPLICATION IN PRECISION AGRICULTURE

For more than 30 years remote sensing has been envisioned as a valuable source of information for crop management. Although much research and development was directed at large-area crop inventory applications of satellite data in the 1970s [12], much less attention has been directed at crop management applications. [13] described the potential for remote sensing in crop management, and stressed that it is critical to provide frequent coverage, rapid data delivery, spatial resolution of 5 to 20 meters, and integration with agronomic and meteorological data into expert systems.

A sequence of remotely sensed images over time can provide information about crop growth and the spatial variation within fields. Detailed spatially distributed multi-temporal information, in visual form, is not readily obtainable from conventional crop management systems or from site-specific crop management methods [14]. Remotely sensed images show spatial and spectral variation resulting from soil and crop characteristics. These images show the state or condition of fields when the images were acquired. One of the most useful aspects of remote sensing is its ability to generate images showing the spatial variation in fields caused by natural and cultural factors. This information is not limited by sampling interval or geostatistical interpolations [15]. Images acquired at different times during a season can be used to determine changes such as growth rates and condition. These data, in turn, can be compared with data from previous years and may be helpful in predicting yield for sustained farm production.

Remote sensing sensors vary in terms of the spatial, spectral, radiometric, and temporal resolution they provide. The size of the pixel used to represent the ground's surface area determines a sensor's spatial resolution. Sensors with tiny footprints and high spatial resolution are more common than those with big footprints and low spatial resolution [16]. The sensor platform rather than the sensor itself might be thought of as having a relationship with temporal resolution. The quantity of bands recorded within a specific range of the electromagnetic spectrum serves as a measure of a sensor's spectral resolution.

Several (10s to 100s) contiguous bands with a narrow width (20 nm) are found in hyperspectral photographs, spaced apart by small wavelength increments [17]. Several vegetation indices, statistical methods, and machine learning techniques, including deep convolutional neural network and random forest, have been used to reduce the dimensionality of hyperspectral data and extract relevant information on crop conditions. In more recent times, estimation of photosynthesis, plant nutrients, and biotic and abiotic stressors like disease and water stress has been done by quantifying solar-induced chlorophyll fluorescence (SIF) from hyperspectral pictures [18].

The majority of publicly accessible satellite products are rather coarse for many PA applications, despite the fact that many contemporary satellites give high spatial (5 m) and temporal (daily) resolution imagery. The right spatiotemporal resolution needed for PA depends on a number of variables, including the size of the field, the management objectives, and the capacity of agricultural equipment to change the rates at which inputs (such as irrigation, fertiliser, pesticides, etc.) are applied. When compared to irrigation and variable rate fertiliser treatments, crop biomass and yield calculations often require higher spatial resolution (1–3 m). Additionally, spatial resolution that is finer than the weed patches is needed for weed mapping and variable herbicide treatment.

The use of UAVs and other sensors mounted on farm equipment (such as spray booms and fertiliser applicators) as well as hand-held, tractor-mounted, and other devices has grown significantly during the past two decades. In-demand data at the spatial scale required for PA operations can be provided by UAVs equipped with multispectral, hyperspectral, and thermal sensors. Due to cloud cover and/or other restrictions/uncertainties related to the sensor platform, getting continuous or frequent satellite scanning during a crop growing season might be challenging. Yet, UAVs can be flown numerous times throughout a growing season to gather the necessary information on a cm-scale. The data at cm-scale required for many field-scale PA applications, such as weed mapping and disease identification, is not typically provided by satellites.

1. Soil mapping can be obtained through the usage of remote sensing satellite imageries. Soil mapping is necessary in Precision Agriculture for the sustainable management of natural resources at all scales, whether local, regional, or

national [19]. The operational success of variable-rate technology depends on maps that show crop conditions such as growth, disease, weeds, and nutrient deficits. Variations in the pace of soil and crop inputs are feasible using VRT. As sensors are installed on satellites, aircraft, or the ground, remotely sensed images from various platforms are used to create these maps.

2. Crop classification is one of the most significant remote sensing applications in precision agriculture. The implementation of additional remote-sensing-based field applications also requires crop classification (e.g., yield estimation, cropping patterns, anomaly detection, etc.). The biophysical characteristics of the Earth's surface and their spatial fluctuation through time are revealed by satellite studies of the planet. This is a fairly comprehensive information source that may be used to determine the types of crops being grown and to keep track of them throughout their growing cycles [20]. The performance of the remote sensing approach may also be impacted by the crop variability in terms of its agroclimatic conditions and field sizes, as it has typically been used in regions with big fields and rather homogenous topography.
3. Remote sensing is a more convenient and effective method for obtaining information about **crop water content** than in situ monitoring. Due to the majority of irrigation field size, which can range from ten to one hundred hectares, it may be prohibitive to even consider taking measurements at ground level [21]. By providing water when a crop needs it or when signs of water stress arise, remote sensing can provide more effective irrigation water management. The crop water stress index (CWSI), which links the temperature of the crop canopy to both the air temperature and the ambient vapour pressure, is the most popular of the several indices designed to estimate crop water status. There have been reports of the use of other indices to determine crop water status, including canopy temperature variability (CTV), temperature stress day (TSD), canopy-air temperature difference (CATD), and stress degree day (SDD).

   Various remote sensing methods are employed to quickly assess **crop nutrition and growth**. Three helpful and frequently used methods are canopy colour analysis, hyper-spectral remote sensing, and visible light remote sensing. Due to the advantages of stable, quick, accessible, and non-destructive results, as well as the strong correlation between the colour parameter of the NRI canopy and the N content of the crop and performance index, these are likely to become hypothetical and non-destructive diagnostic methods for nitrogen crop nutrition in this new era. Moreover, regular monitoring of spatial variation in nutrients, made feasible by remote sensing, can help to uncover seasonal patterns in nutrient distribution across a grower's fields and farm, which can serve to advise more precise long-term fertiliser delivery plans. This can help decrease over-application and the harmful impacts of this practice on the environment while also resulting in extra fertiliser savings [22].
4. Large-scale soil moisture can be estimated using satellite and aircraft remote sensing, which allows us to more accurately simulate the interactions between the land and the atmosphere [23]. The capacity to observe soil moisture at both the global and regional scales using active and passive microwave remote

sensing helps meet the scientific and practical requirements for hydrology. Together with other factors including surface temperature, surface roughness, and vegetation, soil moisture affects the emissive and scattering properties of soil surfaces.

5. The agricultural yield can be forecasted by remote sensing using the amount of light that plants receive. Data from remote sensing are an effective technique for calculating agricultural yield. One can estimate agricultural productivity using the data it provides about the crops that are growing and their surroundings [24]. Crop forecasting, as well as the monitoring and evaluation of droughts, are aided by the integrated use of crop models with geospatial tools and precise observation networks. Previous research indicates that a number of studies have recently used remote sensing indices for crop monitoring and crop production forecasts, including the Enhanced Vegetation Index (EVI), Green Normalized Difference Vegetation Index (GNDVI), and Weighted Difference Vegetation Index (WDVI) [25].
6. Several imaging sources are well-known for having unique spectral, spatial, radioactive, and temporal properties, making them appropriate for various **vegetation mapping** applications. In general, it is necessary to first construct a vegetation classification at either the community or species level in order to map and categorise the plant cover from remote sensing photos [26]. The next step is to find correlations between the plant kinds (species or communities) within this categorisation system and spectral features of remote sensing data that can be clearly seen. At the end of the image interpretation process, also known as image processing, these spectral classifications of the images are converted into kinds of vegetation. Additionally, a variety of vegetation indices, including the soil adjusted vegetation index (SAVI), enhanced vegetation index (EVI), atmospherically resistant vegetation index (ARVI), and normalised differential vegetation index (NDVI), have been used as indirect indicators to gauge and map changes in vegetation cover. Vegetation indices highlight specific aspects of vegetation by combining surface reflectance at various wavelengths.
7. The growth and development of plant roots are significantly influenced by **evapotranspiration (ET),** which is a critical component of energy transmission. In this instance, ET has significant implications for monitoring and assessing droughts across wide areas, especially irrigated agricultural areas [27]. Traditionally, ET was determined indirectly from environmental factors at meteorological stations, such as solar radiation, air temperature, wind speed, and relative humidity, because direct measurement of ET is time-consuming, expensive, and complicated. An alternative method involves estimating ET at local, regional, and global scales by fusing surface parameters obtained from remote sensing data with vegetation traits and surface meteorological factors. Multispectral data with various temporal and spatial resolutions can be provided by various remote sensing sensors on various platforms, including Landsat, the Moderate Resolution Imaging Spectroradiometer, the Advanced Very High-Resolution Radiometer, the Visible Infrared Imaging

Radiometer Suite, and the Advanced Spaceborne Thermal Emission and Reflection Radiometer. For estimating ET using remotely sensed data, a number of models have been proposed, including the Surface Energy Balance Algorithm for Land Model (SEBAL), Mapping Evapotranspiration at High Resolution with Internalized Calibration (METRIC), the Simplified Surface Energy Balance Index (SSEBI), the Surface Energy Balance System Model (SEBSM), and the Surface Energy Balance Algorithm for Land-Improved Model (SEBALI). In various climatic regions of the world, remote sensing has been used to estimate ET with effectiveness.

8.3 GEOGRAPHIC INFORMATION SYSTEM (GIS)

Available GIS software ranges from simple map display systems to fully functioning systems capable of analysing and integrating complex spatial databases. Some data can be stored as polygons within which attributes like soil types are considered to be homogeneous. Data can also be stored in a uniform array of grid cells or pixels with homogeneous attributes, which is the format used for remote sensing images. Most fully functioning GIS programs today can be converted between these formats, which have made it easier to combine data from different sources.

Among the most important roles for GIS are the database functions for farm record keeping and for comparing management decisions, yields, pest activity, groundwater quality, and other concerns related to past and current practices. GIS can store farm records of inputs and outputs in a spatial array. A key to realising the promise of a dynamic GIS will be the development of connections between the relational database and the decision support system. A disadvantage of the current generation of geographic information systems is the complexity of the software and the steep learning curve involved in using and interpreting spatial data in a valid and robust situation. This is rapidly changing as several vendors are developing fully functional GIS programs intended for use on PCs. There is an urgent need to make fully functional GIS easier for non-specialists to learn and use in order to transfer this technology to the agricultural community [28].

8.4 GLOBAL POSITIONING SYSTEM (GPS)

The Global Positioning System (GPS) is a system of satellites emitting electronic signals that can be received by mobile field instruments sensitive to the transmitting frequency. Positioning is achieved through the use of simultaneously received satellite transmissions from four or more satellites above the horizon. With a constellation of 24 satellites, any location on Earth can have four or more satellites in view for 24 hours each day. By referencing the satellite's exact location and the time the signal takes to travel between the transmitter and the receiver, the location of the receiver can be determined by triangulation [29].

Use of the GPS receiver allows latitude and longitude coordinate information to be associated with data obtained from a specific site on the field. The GPS can also be used to provide navigational guidance, enabling a producer to revisit a spot in the field and check the efficacy of management decisions. The GPS is an essential field component for most mapping-based managements and other measurements of field characteristics that would be used to determine product application maps.

The positions provided by GPS receivers currently are not sufficiently accurate for dynamic real-time uses. However, technical solutions are available to improve the positioning accuracy. A technique known as differential correction is widely used to remove the effects of the error sources.

8.5 VARIABLE RATE TECHNOLOGY (VRT)

Variable rate technology is the most advanced component of PF technologies, and provides "on-the-fly" delivery of field inputs. A GPS receiver is mounted on a truck so that a field location can easily be recognized. An in-vehicle computer, which contains the input recommendation maps, controls the distribution valves to provide a suitable input mix by comparing to the positional information received from the GPS receiver. Current commercial VRT systems are either map-based or sensor-based [30]. The map based VRT systems require a GPS/DGPS geo-referenced location and a command unit that stores a plan of the desired application rates for each field location. The sensor-based VRT systems do not require a geo-referenced location but include a dynamic control unit, which specifies application through real-time analysis of soil and/or crop sensor measurements for each field location. New VRT systems like the manure applicator may soon enable precise application of manure in cropping systems [11].

There are two methods of VRT. The first method, map-based, includes the following steps: grid sampling a field, performing laboratory analyses of the soil samples, generating a site-specific map of the properties and finally using this map to control a variable-rate applicator. During the sampling and application steps, a positioning system, usually DGPS (Differential Global Positioning System) is used to identify the current location in the field.

The second method, sensor-based, utilizes real-time sensors and feedback control to measure the desired properties on-the-go, usually soil properties or crop characteristics, and immediately use this signal to control the variable-rate applicator.

8.6 CONCLUSION

Remote sensing was used in almost every element of PA from land preparation to harvesting. The face of PA has altered as a result of the quantity of high spatial resolution, multi-temporal satellite data, affordable unmanned aerial vehicles and commercially available ground-based proximity sensors. The prospective uses of remote sensing in PA have been examined using a variety of cutting-edge techniques, including empirical, regression, and various machine learning approaches. Similarly, many vegetation indices have been developed and tested for their ability to help PA operations, including variable fertilizer management, irrigation scheduling, disease control, weed mapping, and yield forecasting. However, many challenges need to be addressed before the remote sensing technologies can potentially see a large-scale adoption in commercial and non-commercial agriculture.

Although the majority of satellite data are freely accessible, processing them for practical purposes may necessitate a substantial amount of technological know-how and skill. Furthermore, many PA procedures, like the management of weeds and diseases, require data with fine spatial resolution (cm-scale) and high spectral and temporal (e.g., daily) precision. Most satellite data that is made available to the public does not adhere to these standards. Moreover, many satellite images might not be acceptable for usage on cloudy days or when there is irregular or changeable irradiance from the sun. The accuracy of techniques based on remote sensing (satellite, aerial, and UAV) data depends on a number of variables, such as image resolution (spatial, spectral, and temporal); atmospheric, climatic, and weather conditions; crop and field conditions (such as growth stage, land cover); and the analysis technique (e.g., regression-based, machine learning, physically based modelling). The development of tools and frameworks that can make it easier for end users to access satellite data for real-time applications still faces significant obstacles and limitations. More use of remote sensing data in commercial and non-commercial PA operations is likely to result from the development of precise, user-friendly systems.

REFERENCES

1. Robertson, G. P., & Swinton, S. M. (2005). Reconciling agricultural productivity and environmental integrity: A grand challenge for agriculture. *Frontiers in Ecology and the Environment*, **3**(1), 38–46.
2. Gomiero, T., Pimentel, D., & Paoletti, M. G. (2011). Is there a need for a more sustainable agriculture? *Critical Reviews in Plant Sciences*, **30**(1–2), 6–23.
3. Grisso, R. D., Alley, M. M., Thomason, W. E., Holshouser, D. L., & Roberson, G. T. (2011). Precision farming tools: Variable-rate application.
4. Rigney, D. J. (2011). Precision agricultural tools for measuring water use efficiency in sorghum. *The Precision Agriculture Research Group. The University of New England Armidale The Australia and New Zealand Cooperative Research Centre for Spatial Information.*
5. Balafoutis, A., Beck, B., Fountas, S., Vangeyte, J., Van der Wal, T., Soto, I., ... & Eory, V. (2017). Precision agriculture technologies positively contributing to GHG emissions mitigation, farm productivity and economics. *Sustainability*, **9**(8), 1339.
6. Pinter Jr, P. J., Hatfield, J. L., Schepers, J. S., Barnes, E. M., Moran, M. S., Daughtry, C. S., & Upchurch, D. R. (2003). *Remote Sensing for Crop Management. Photogrammetric Engineering & Remote Sensing,* 69(6), 647–664.
7. Diacono, M., Rubino, P., & Montemurro, F. (2013). Precision nitrogen management of wheat. A review. *Agronomy for Sustainable Development*, **33**, 219–241.
8. Koch, B., Khosla, R., Frasier, W. M., Westfall, D. G., & Inman, D. (2004). Economic feasibility of variable-rate nitrogen application utilizing site-specific management zones. *Agronomy Journal*, 96(6), 1572–1580.
9. Khanal, S., Fulton, J., & Shearer, S. (2017). An overview of current and potential applications of thermal remote sensing in precision agriculture. *Computers and Electronics in Agriculture*, **139**, 22–32.
10. Shanmugapriya, P., Rathika, S., Ramesh, T., & Janaki, P. (2019). Applications of remote sensing in agriculture – A Review. *International Journal of Current Microbiology and Applied Sciences*, **8**(01), 2270–2283.

11. Khanna, M. (2021). Digital transformation of the agricultural sector: Pathways, drivers and policy implications. *Applied Economic Perspectives and Policy*, **43**(4), 1221–1242.
12. Navalgund, R. R., Jayaraman, V., & Roy, P. S. (2007). Remote sensing applications: An overview. *Current Science*, 1747–1766.
13. Jackson, R. D. (1984, October). Remote sensing of vegetation characteristics for farm management. In *Remote Sensing: Critical Review of Technology* (Vol. 475, pp. 81–97). SPIE.
14. Corwin, D. L., & Lesch, S. M. (2005). Apparent soil electrical conductivity measurements in agriculture. *Computers and Electronics in Agriculture*, **46**(1–3), 11–43.
15. Van Westen, C. J. (2013). Remote sensing and GIS for natural hazards assessment and disaster risk management. *Treatise on Geomorphology*, **3**(15), 259–298.
16. Aasen, H., Honkavaara, E., Lucieer, A., & Zarco-Tejada, P. J. (2018). Quantitative remote sensing at ultra-high resolution with UAV spectroscopy: A review of sensor technology, measurement procedures, and data correction workflows. *Remote Sensing*, **10**(7), 1091.
17. Kim, M. S., Chao, K., Chan, D. E., Jun, W., Lefcourt, A. M., Delwiche, S. R., ... & Lee, K. (2011). Line-scan hyperspectral imaging platform for agro-food safety and quality evaluation: System enhancement and characterization. *Transactions of the ASABE*, **54**(2), 703–711.
18. Hernández-Clemente, R., Hornero, A., Mottus, M., Peñuelas, J., González-Dugo, V., Jiménez, J. C., ... & Zarco-Tejada, P. J. (2019). Early diagnosis of vegetation health from high-resolution hyperspectral and thermal imagery: Lessons learned from empirical relationships and radiative transfer modelling. *Current Forestry Reports*, **5**, 169–183.
19. Huang, Y., Chen, Z. X., Tao, Y. U., Huang, X. Z., & Gu, X. F. (2018). Agricultural remote sensing big data: Management and applications. *Journal of Integrative Agriculture*, **17**(9), 1915–1931.
20. Atzberger, C. (2013). Advances in remote sensing of agriculture: Context description, existing operational monitoring systems and major information needs. *Remote Sensing*, **5**(2), 949–981.
21. Mulla, D., & Khosla, R. (2016). Historical evolution and recent advances in precision farming. *Soil-specific Farming Precision Agriculture*, **1**, 35.
22. Ayaz, M., Ammad-Uddin, M., Sharif, Z., Mansour, A., & Aggoune, E. H. M. (2019). Internet-of-Things (IoT)-based smart agriculture: Toward making the fields talk. *IEEE Access*, **7**, 129551–129583.
23. Janani, N., Kannan, B., Nagarajan, K., Thiyagarajan, G., & Duraisamy, M. R. (2023). Soil moisture mapping for different land-use patterns of lower Bhavani river basin using vegetative index and land surface temperature. *Environment, Development and Sustainability*, **26**(2), 4533–4549.
24. Lobell, D. B., Asner, G. P., Ortiz-Monasterio, J. I., & Benning, T. L. (2003). Remote sensing of regional crop production in the Yaqui Valley, Mexico: estimates and uncertainties. *Agriculture, Ecosystems & Environment*, **94**(2), 205–220.
25. El-Tantawi, A. M., Bao, A., Chang, C., & Liu, Y. (2019). Monitoring and predicting land use/cover changes in the Aksu-Tarim River Basin, Xinjiang-China (1990–2030). *Environmental Monitoring and Assessment*, **191**, 1–18.
26. Jones, H. G., & Vaughan, R. A. (2010). *Remote Sensing of Vegetation: Principles, Techniques, and Applications*. Oxford University Press, USA.

27. Karishma, C. G., Kannan, B., Nagarajan, K., Panneerselvam, S., & Pazhanivelan, S. (2022). Spatial and temporal estimation of actual evapotranspiration of lower Bhavani basin, Tamil Nadu using Surface Energy Balance Algorithm for Land Model. *Journal of Applied and Natural Science*, **14**(2), 566–574.
28. Huang, B., Jiang, B., & Li, H. (2001). An integration of GIS, virtual reality and the Internet for visualization, analysis and exploration of spatial data. *International Journal of Geographical Information Science*, **15**(5), 439–456.
29. Bevis, M., Businger, S., Herring, T. A., Rocken, C., Anthes, R. A., & Ware, R. H. (1992). GPS meteorology: Remote sensing of atmospheric water vapor using the global positioning system. *Journal of Geophysical Research: Atmospheres*, **97**(D14), 15787–15801.
30. Sahoo, R. N. (2011). Precision farming: Concepts, limitations, and opportunities in Indian agriculture. *Resource Conserving Techniques in Crop Production*, 439–450.

9 Harnessing Advanced Farming with Unmanned Aerial Vehicles

Mohammad Shahnawaz Shaikh

9.1 INTRODUCTION

Many countries with heavy reliance on agriculture are adopting technology to optimize farming practices. Drones have traditionally been considered expensive toys, but the agricultural industry has overlooked their potential. Developed nations have already embraced the use of UAVs in advanced agriculture, employing techniques such as photogrammetry and remote sensing. UAVs offer rapid and efficient solutions that can alleviate the workload of farmers. In order to monitor crops, they are outfitted with sensors, cameras, and pesticide sprayers. An in-depth examination of UAVs in precision agriculture involves analyzing their suitability for tasks such as crop monitoring, estimation of crop height, pesticide spraying, and soil and field analysis [1]. However, the successful implementation of these UAVs heavily relies on critical factors including weight, flight range, payload capacity, configuration, and cost. With the use of specialized software, drones can function independently. This software enables the design of flight plans, GPS system deployment, and the input of several parameters including speed, altitude, Region of Interest (ROI), geo-fence, and fail-safe modes. Due to their excellent spatial resolution, rapid turnaround times, low operating costs, and simplicity of activation, drones are favored over full-sized aircraft. These qualities are critical in precision agriculture, where it's necessary to monitor and evaluate broad areas quickly [2]. Compact cameras and other sensors like sonar and infrared have shrunk in size, making aerial vehicle use feasible. The goal of this chapter is to give a thorough review of how Unmanned Aerial Vehicles (UAVs), also referred to as drones, are being used in the agricultural industry. In particular, how they are used to track and assess plant stresses like disease, drought, nutrient shortages, pests, and weeds is the main focus. Agriculture operations include a lot of monitoring of crops for pests, nutrients, diseases, water stress, and general plant health. Traditionally, this activity has involved surveying the crops on the ground or from the air [3].

Nevertheless, there are financial and accessibility constraints with these techniques. On the other hand, imagery generated by light aircraft is more affordable, delivers greater resolution, and gives more recent data. Even yet, the cost per

DOI: 10.1201/9781003487005-9

acre remains somewhat high. Alternatively, centimeter-level resolution temporal and geographic data collection can be achieved with small unmanned aerial vehicles (UAVs) or unmanned aircraft systems (UAS). The crop fields can be thoroughly covered by these UAVs since they can reliably fly at a certain height following predefined paths. Cloud cover has no discernible impact on UAV image acquisition procedures. The use of UAS in agriculture has a number of restrictions and difficulties despite its many benefits. The key UAS technology and uses in the agricultural sector are outlined in this chapter, along with a discussion of the difficulties in implementing them [4–5].

9.2 UNMANNED AERIAL VEHICLE (UAV)

Unmanned Aerial Vehicles (UAVs) refer to aircraft that are capable of autonomous flight, eliminating the need for a human pilot, and are operated through radio communication. Among the various types of UAVs, multi-rotors stand out as they are categorized based on the number of rotors they possess. Fixed-wing UAVs (depicted in Figure 9.1(a)) exhibit distinct design characteristics in comparison to multi-rotors, with their aerodynamic wing configuration facilitating smooth gliding. Single-rotor helicopters (Figure 9.1(b)) consist of one large rotor positioned at the top and a smaller one at the tail of the UAV. Multi-rotors, such as the quadcopter (Fig. 9.1(c)), Hexacopter (Fig. 9.1(d)), and octocopter (Fig. 9.1(e)), rely on four, six, or eight rotors, respectively, for lift and propulsion.

A quadcopter is a unique type of unmanned aerial vehicle (UAV) that consists of four rotors. These rotors are responsible for generating the lift required for the quadcopter to fly. Two of the rotors in this layout rotate in a clockwise (CW) direction, while the other two rotate counter-clockwise (CCW).

The movement of the quadcopter is facilitated by its ability to pitch (move backward and forward), roll (move left and right), and yaw (rotate clockwise and counter-clockwise) around its axis. The quadcopter can be configured in two different ways: the plus (+) model and the cross (X) model, as depicted in Figures 9.2(a) and 9.2(b) respectively. The cross model, which is shown in Figure 9.2(b), is particularly popular and offers greater stability when compared to the plus model.

a b c d e

FIGURE 9.1 Types of UAV or Drone, (a) Fixed Wing, (b) Single Rotor, (c) Quadcopter, (d) Hexacopter, (e) Octo Copter.

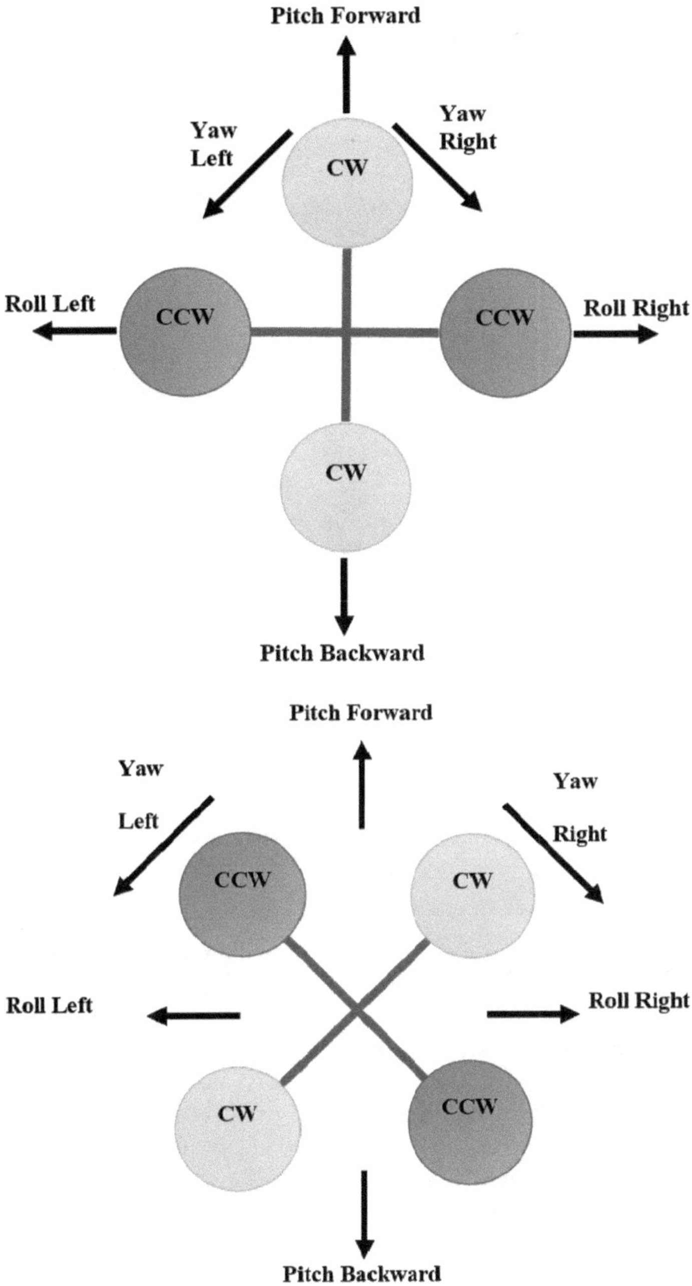

FIGURE 9.2 Quadcopter Configuration: (a) Plus Configuration, (b) Cross Configuration.

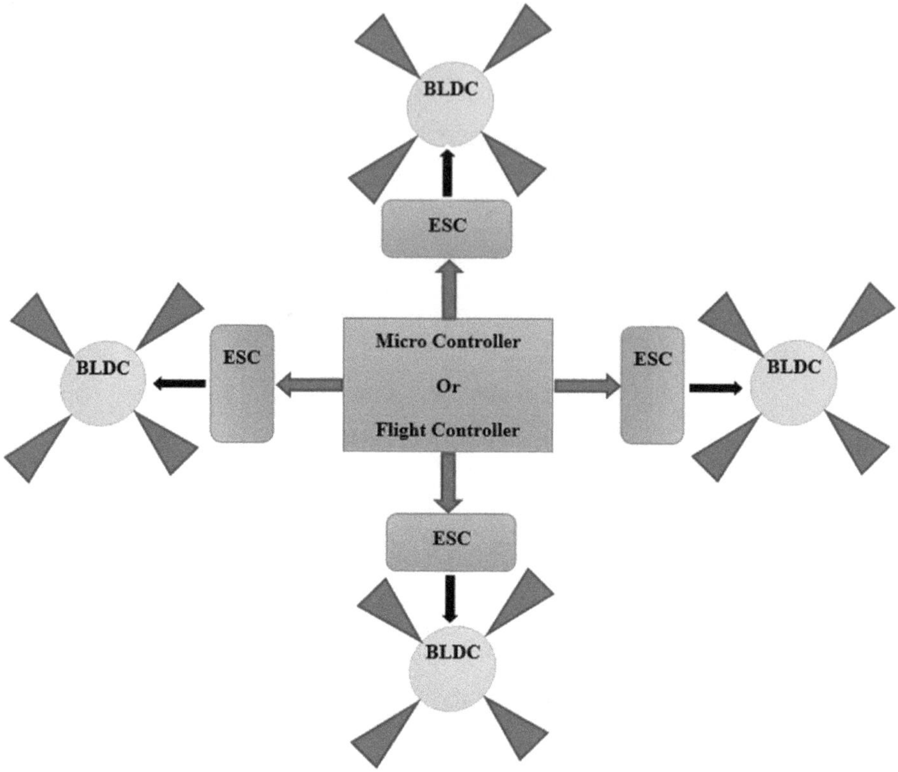

FIGURE 9.3 Block Diagram of Drone.

9.3 METHODOLOGY

The flight controller serves as the primary board within the unmanned aerial vehicle (UAV), encompassing advanced firmware that assumes responsibility for the actual flight. Simultaneously, the flight controller manages numerous aspects of the UAV's operation. It is constructed with a microcontroller that establishes communication with the four brushless motors. In accordance with the UAV configuration model, the brushless direct current (BLDC) motors are attached to the rotors of the UAV [6]. These BLDC motors are controlled by Electronic Speed Controllers, or ESCs. In addition, a transmitter and receiver of UAV controls by a radio channel. Every remote control (RC) transmitter has many channels, allowing for the independent control of different UAV-related tasks. Figure 9.3 illustrates a representative block diagram of a drone [7].

9.4 APPLICATIONS OF DRONES IN AGRICULTURE

Drones are improving farming in many ways. Major farming operations have become easier, faster and more cost effective with fine monitoring due to use of drones. Figure 9.4 shows major drones applications in advanced and smart farming.

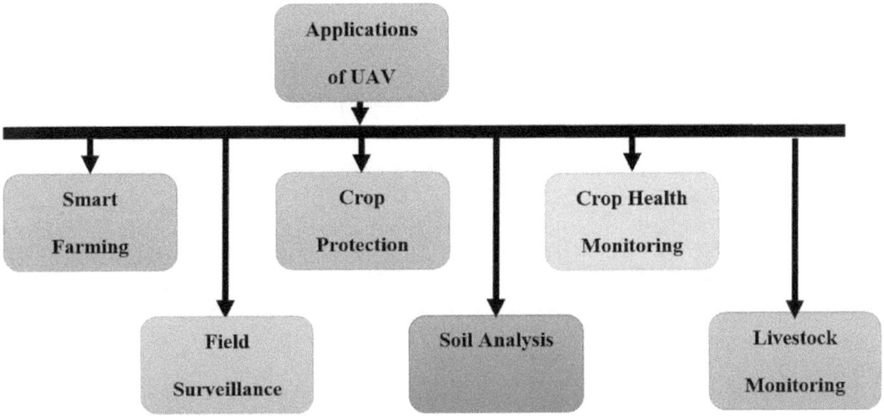

FIGURE 9.4 Applications of Drones.

9.4.1 Smart Farming

9.4.1.1 Planting

Planting seeds using unmanned aerial vehicles (UAVs) offers a potential solution to the challenges posed by manual labor in terms of both time and accessibility. Geographical barriers such as rivers, hills, and ravines, which pose safety risks for workers, can be easily overcome by drones. The performance of the seed planting mechanism is typically influenced by factors such as the volume of the canister, battery life, and motor specifications. While some researchers focus solely on designing the mechanism and implementing it on commercially available drones, others have also developed drones that are integrated with the seed planting mechanism. Quadcopter drones are available with a seed dropping mechanism specifically for the purpose of re-vegetation. This mechanism first acquires an aerial image of the targeted field and then designs the seed dispersing process accordingly. Upon reaching the predetermined positions, the seeds are dispersed [8–9].

9.4.1.2 Sprinkling System

The World Health Organization (WHO) has estimated that there are in excess of one million instances of pesticide exposure annually. These occurrences result in more than 100,000 deaths each year, particularly in developing nations. This is primarily due to the use of pesticides by human beings. The impact of these chemicals on the human nervous system can lead to various disorders within the body. To address this issue, a remotely controlled Unmanned Aerial Vehicle (UAV) is employed to administer pesticide and fertilizer, thus minimizing human exposure to toxic substances [10].

As shown in Figure 9.5, the sprinkling system, in general, is connected to the lower area of the UAV. It possesses a nozzle located beneath the pesticide tank, which is responsible for dispensing the pesticide downstream. This system is composed of two modules: the sprinkling system itself and the Controller. The sprinkling system is equipped with a nozzle that is used for spraying the spraying content, be it pesticides

FIGURE 9.5 Sprinkling and Spraying by Drone.

or fertilizers. On the other hand, the Controller is responsible for activating the sprayer's nozzle. To facilitate the flow of the pesticide through the nozzle, a pressure pump is an integral component of the sprinkler system. In order to meet the required pressure, a motor driver integrated circuit is employed to regulate the pump [11].

9.4.1.3 Spraying

Agricultural productivity relies on the use of materials for production and protection. To promote crop growth and eliminate insects, the application of fertilizers and chemicals is often necessary. The utilization of drones provides a potential solution for the precise and efficient spraying of these chemicals, taking into consideration the spatial variability of the crops and field. The quantity of chemicals sprayed can be adjusted based on the specific conditions of the crop or the severity of the insect-pest attack. By integrating sprayer systems with UAVs, a platform for pest management and vector control can be established, particularly for large crop fields. In order to cover extensive areas, heavy lift UAVs are necessary. The low-cost and lightweight compact quadcopter (QC) system can be used for both indoor and outdoor crops. The quadcopter has the capability of autonomous flight for the purpose of spraying pesticides and fertilizers, which can be controlled through an android device [12]. The quadcopter and android device are able to achieve real-time communication using a Bluetooth device. This system fulfils the objective of precise and site-specific application. The utilization of this particular system serves the purpose of mitigating issues related to agriculture fields, while simultaneously enhancing agricultural yield [13].

The efficiency of the spraying system, which is affixed to the Unmanned Aerial Vehicle (UAV), experiences improvement through the implementation of the Pulse Width Modulation (PWM) controller in the application of pesticides. A quadcopter-based automated pesticide sprayer, integrated with a blimp and known as the Aerial Automated Pesticide Sprayer (AAPS), has been devised for the purpose of pesticide spraying, utilizing the GPS coordinates in an environment characterized by lower altitudes. The utilization of drones to administer chemicals in agriculture has the potential to enhance the efficiency of these substances, resulting in reduced negative effects on the environment, specifically in terms of soil and water pollution. Consequently, the adoption of this technology can contribute to the establishment of sustainable agricultural practices. Drones have the advantage of being able to distribute chemicals at a faster pace compared to alternative methods. Furthermore, their usage can lead to a reduction in the amount of chemicals required, thereby resulting in cost savings. In agricultural seasons, the application of fertilizers and micronutrients is a crucial economic input. Traditionally, these substances are applied using ground equipment or manned aircraft. The latter option is generally favored by producers with extensive land holdings [14]. However, the inability to accurately adjust application rates due to changing wind conditions and aircraft elevation limits the precision of this method. To overcome this challenge, ground equipment application is employed alongside aerial spraying to ensure consistent crop nutrient levels throughout the irrigation season.

Furthermore, this technology allows for precise crop dusting, surpassing the capabilities of traditional tractors. By adopting this method, costs are reduced, and the risk of pesticide exposure to manual laborers is mitigated. The utilization of drones for pesticide application is advantageous in all scenarios, particularly in areas where labor is scarce. Additionally, this approach minimizes environmental pollution when sprayed from lower altitudes and presents significant potential for improving pest management in both small and large-scale crop fields through highly accurate site-specific applications [15].

9.4.2 Field Surveillance

Field surveillance is one of the major and useful applications of drones. It is very helpful for monitoring the field as well as monitoring of resources available on a field. Some major use cases discussed as follows:

9.4.2.1 Evapotranspiration (ET) Estimation by Drone

Evapotranspiration or ET is a major phenomenon in which soil evaporation and plant transpiration work together to move water from the land to the atmosphere. Professionals in the fields of hydrology, agriculture, and water management rely on estimates of potential ET to inform their work. Due to the pressing issues of water scarcity, population growth, and climate change, the estimation of evapotranspiration has become a crucial area of research in agriculture. In this regard, unmanned aerial vehicles (UAVs) have emerged as valuable tools, serving various research purposes for ET estimation. These UAVs can be classified into three main platforms: aircraft, fixed-wings, and quadcopters [16]. While aircrafts tend to be costly, they offer extended flight duration and the capability to carry heavy sensors. On the other hand, fixed-wings and quadcopters are relatively more affordable alternatives. Fixed-wings can typically fly for about two hours, making them suitable for large-scale field operations. Quadcopters, on the other hand, are designed for short flight missions in small-scale fields, with a flight time of approximately 30 minutes. Drone image processing and flight path planning are two new research issues that come with using UAVs as remote sensing platforms. For instance, two source energy balance models have been used to estimate ET using a fixed-wing UAV fitted with thermal data collection capabilities [17].

9.4.2.2 Water Stress Monitoring

Characterizing the impact of water stress on crops is a multifaceted endeavor due to the intricate interplay between drought effects and various influencing factors. The utilization of thermal images to derive variables often relies on minute temperature fluctuations to identify stresses and other occurrences. Thus, thresholds and regression equations that are developed under particular conditions are usually not applicable to even slightly different situations. For example, inherently differing stomatal conductance and transpiration rates can result in differences in canopy temperatures between genotypes of a given crop growing under the same conditions [18–19].

Various types of sensors and models were employed by researchers to detect water stresses. The utilization of multispectral or hyperspectral images, as well as vegetation indices such as NDVI and GNDVI, represents spectral transformations aimed at highlighting specific vegetation characteristics. Moreover, the photochemical reflectance index (PRI) combined with the use of multispectral or hyperspectral pictures provides a reflectance measurement that is responsive to changes in the carotenoid pigments present in leaves. Additionally, some studies that directly use the canopy temperature also use thermal infrared images to calculate the temperature differential (Tc–Ta) between the air temperature and the canopy temperature. Furthermore, the crop water stress index (CWSI) and the deployment of thermal infrared images depend on the difference between canopy and air temperatures (Tc–Ta), which is then normalized by the vapor pressure deficit (VPD) [20].

The reason behind this phenomenon is that when plants experience water stress, it leads to a reduction in stomatal conductance and a decrease in heat dissipation, resulting in a noticeable rise in canopy temperature. The utilization of Red-Green-Blue (RGB) images has been limited, typically in conjunction with multispectral or thermal images, in order to calculate hybrid variables like the Water Deficit Index (WDI). Additionally, the sporadic application of chlorophyll fluorescence, determined through narrow-band multispectral images, has also been employed for the purpose of detecting and monitoring water stress [21–22].

9.4.3 Crop Protection

Crop protection has become very effective with the involvement of drones in agriculture. It helps the farmers to increase their revenue by reducing the losses due to non-effective care of crop.

9.4.3.1 Weed Control

Unwanted plants like weeds can spread throughout agricultural crops and cause a host of problems. These plants compete with one another for scarce resources like water and space, which has a negative impact on crop development and yields. Weeds have a substantial negative impact on agricultural yield losses, affecting a wide range of crops including potatoes, rice, wheat, corn, sugarcane, vegetables, and jute. The most common approach to controlling weeds is to employ pesticides in order to tackle this issue. Conventional farming entails removing weeds after they have emerged, but the standard method of managing weeds is to apply the same amount of herbicide to the entire field, including the regions that are weed-free. However, overuse of herbicides might result in the emergence of weeds resistant to them, which will ultimately hinder crop growth and productivity. Consequently, the application of hyperspectral images has been employed to differentiate between the spectral signatures of different weed species that possess varying levels of resistance to glyphosate. Additionally, RGB sensors have been utilized to categorize various weed species [22–23].

Unmanned aerial vehicles equipped with hyperspectral sensors can monitor the presence of unwanted vegetation. The monitoring was conducted by assessing the chlorophyll content and leaf density of the plant canopy. It's also important to emphasize how dangerous this technology is for causing pollution in the environment. Weed management techniques tailored to individual sites have been put into place to solve the aforementioned issues. Creating a precise map that shows the amount of weed coverage is essential to getting the intended result. This would enable precise application of herbicides in areas requiring treatment. The use of drones allows for comprehensive data collection and analysis across the entire field, facilitating the creation of a detailed weed cover map that pinpoints the specific locations in need of chemical intervention [24].

The application of agro-drones for the purpose of spraying weedicide proves to be beneficial in the control of weeds during both the pre-emergence and post-emergence stages. This spraying mechanism is capable of functioning effectively in various field conditions, including those that are muddy, infested with weeds, or plagued by insects. Furthermore, it can operate efficiently in both sunny and drizzling weather conditions. The utilization of drones for weedicide application not only ensures efficacy but also optimizes the usage of this weed control substance. Additionally, the simplicity of usage, portability, and ease of maintenance contribute to its practicality. Moreover, the ability to remotely operate these drones ensures the utmost safety for one's health [25–26].

9.4.4 Soil Analysis

Soil analysis is a crucial technique to improve crop production from the fields. Perfect analysis of soil helps the farmer to decide which type of crop to be processed in a specific season.

9.4.4.1 Soil and Field Analysis

At each phase of the crop cycle, drones can be employed to acquire valuable information. By acquiring three-dimensional maps of the current soil conditions, it will be possible to assess the potential soil quality, regulate nutrient management, and identify areas of soil degradation. This data can aid farmers in determining the most efficient strategies for planting, crop management, soil maintenance, and other related aspects. Consistent monitoring can enhance the optimization of water resources and the effective control of crop nutrient levels [27].

9.4.4.2 Mapping and Soil Analysis

The association between mapping and soil analysis is intricately intertwined, as the outcomes derived from aerial mapping can be utilized to further assess the state of the soil and approximate the potential crop yield. The process of mapping can be carried out manually or automatically through the utilization of drones, adhering to a predetermined route. Consequently, the camera equipped on the drone captures land images along this designated path, subsequently undergoing processing through specialized software. Numerous research endeavors have been undertaken in pursuit of this particular application, focusing on the analysis of soil, mapping techniques, or a combination of both [28].

9.4.5 Crop Health Monitoring

Continuous crop health monitoring is required to treat crop disease at the correct time in the correct way. Healthy crops generate healthy revenue.

9.4.5.1 Health Monitoring

Monitoring the condition of crops, particularly during the different stages of growth, is of great importance to farmers. The data obtained from this monitoring process will have a significant impact on the decision-making process regarding timely interventions that farmers need to implement in order to ensure an optimal yield at the end of the season. The utilization of cameras is crucial in providing the necessary information on the condition of the crops, and the most commonly used types of cameras include digital, thermal, and multispectral. A thermal camera, which is equipped with an infrared sensor, is capable of generating a thermal map containing temperature data pertaining to the crops. On the other hand, digital cameras can process digital images in order to extract color information based on the red-green-blue (RGB) model, while multispectral cameras can capture both visible and invisible images of the crops by utilizing various wavelengths of light. These images are then combined to create an accurate mosaic [29].

Drones can potentially be employed to observe crop conditions over the course of the crop season in order to enable timely and demand-driven action. Using a range of visible, near-infrared, and thermal infrared sensors makes it possible to compute various multispectral indices based on the wavelength-dependent patterns of reflection. The circumstances of crops, including but not limited to water stress, nutritional stress, insect-pest attack, illnesses, and more, can then be evaluated using these indices. The drones' sensors function as an early detection tool by enabling the identification of illnesses or deficits even before to the appearance of symptoms [30].

Drones can therefore be used as an early warning system to enable timely intervention by putting corrective measures in place based on the intensity of the stress. With the use of thermal and multispectral sensors to capture the reflectance of the vegetation canopy, unmanned aerial vehicles (UAVs) are able to fly over large fields in one flight. These cameras are installed on the underside of the quadcopter, and they take one picture per second. The picture is then saved in memory and sent via telemetry to the ground station. An equation representing the Normalized Difference Vegetation Index (NDVI) was used as the geographic indicator for the interpretation of the data obtained through telemetry from the multispectral camera [31–32].

$$NDVI = \frac{R_{NIR} - R_{RED}}{R_{NIR} + R_{RED}}$$

R_{NIR} = Reflectance of the near infrared band. RRED = Reflectance of the red band.
R_{RED} = Reflectance of the red band.

The Normalized Difference Vegetation Index (NDVI) is a simple metric used to measure how vibrant green vegetation is. The basic idea is that red and blue wavelengths are absorbed by chlorophyll, whereas near infrared light (NIR) at about 750 nm is robustly reflected by it. It is because of this powerful chlorophyll reflection that plants look green to the naked eye. But in NIR, the reflection is much more noticeable, and this feature is essential for supplying precise data for analysis. The ensuing computations produce values between -1 and +1, where values near 0 indicate that there is no vegetation on the crop, while values near +1 (between 0.8 and 0.9) indicate that there is the highest density of green leaves on the crop [33–34].

These results make it simple for farmers to evaluate the health of their crops and adjust their monitoring methods accordingly. Furthermore, these findings help farmers pinpoint the precise locations that need pesticide application. Throughout the growing season, drones can be used to monitor crop conditions and provide timely, customized treatments. It is possible that yield loss can be avoided with this prompt and proper approach. Additionally, this technology eliminates the need for farmers to visually monitor their crops, which is very helpful in remote areas like mountainous regions. Farmers can also effectively monitor trees and tall crops, which would be difficult to personally scout [35–36].

9.4.5.2 Nutrient Status and Deficiency Monitoring

Plants require the appropriate quantities of nutrients to flourish and yield robustly. Optimal levels of nitrogen are necessary for promoting vegetation and foliage growth, while adequate levels of phosphorus are crucial for enhancing the development of roots and stems. In addition, suitable levels of potassium are indispensable for bolstering disease resistance and ensuring a superior crop quality. In the absence of any of these essential nutrients in the soil, plants will undergo stress and struggle to thrive [37]. The utilization of NDVI mosaics facilitates the precise identification of stressed or struggling areas within the crops, enabling targeted intervention in these specific zones. Moreover, the NIR/multispectral imagery captured by UAVs allows for the early detection of these management zones, long before they become perceptible to the naked eye. Consequently, these management zones can be addressed prior to any adverse impact on crop growth and yield. Unmanned aerial vehicles (UAVs) equipped with specialized camera sensors, such as optical and thermal cameras, as well as specific optical filters like Red Edge or hyperspectral cameras, can be utilized to estimate crop nutrient status. This information can directly assist producers and agronomist consultants in making appropriate application rate recommendations for the entire field [38–39].

Currently, the most popular method for assessing nutritional status is visual, using plant color guides, which makes it impossible to make precise quantitative assessments. Laboratory leaf analyses are necessary for more precise assessments, but they take time and require specialized skills to interpret data correctly. Alternative indirect approaches exist for some nutrients, such as the prediction of nitrogen levels using the chlorophyll meter (also known as soil-plant analyses development, or SPAD), albeit the procedure is time-consuming and the results are not always accurate. Consequently, a great deal of work has gone into creating novel techniques for determining and estimating nutritional issues in plants. Because nitrogen is linked to yield and biomass, it is a nutrient that has been studied in great detail [39–40].

Sodium and potassium have also drawn considerable attention. Utilizing multispectral pictures is the primary method for extracting relevant features and indices. Nevertheless, RGB and hyperspectral images are also regularly adopted. Additionally, studies have looked into data fusion methods that integrate two or even three different kinds of sensors, like thermal, RGB, and multispectral sensors [40].

9.4.5.3 Crop Diseases Monitoring

Crop diseases can cause extensive damage and can be classified into three main types: fungal, bacterial, and viral. The utilization of unmanned aerial vehicles (UAVs) equipped with infrared cameras enables the visualization of the internal condition of plants, presenting a clear representation of their health status. By identifying infections in their early stages, farmers can implement preventive measures, such as removing the infected plant, to prevent the spread of the infection to nearby plants. Therefore, image-based tools, particularly those facilitated by UAVs, play a pivotal role in the identification and detection of plant diseases, especially in situations where human assessment is impractical, unreliable, or unavailable. The gathering of information about the areas being studied has predominantly relied on RGB and multispectral images, although experiments have also been conducted with hyperspectral and thermal images [40].

9.4.6 Livestock Monitoring

The use of drones for livestock monitoring has garnered significant attention from farmers. The drone-captured images of the livestock can be analyzed to determine their quantity and behavior, which is then transmitted in real-time to monitoring devices for the farmers to respond accordingly. This approach proves highly effective, particularly in the case of monitoring livestock on vast agricultural lands and tracking human poachers. However, a common issue encountered in this application is the misclassification of livestock due to the similarity in physical features between certain species, resulting in inaccurate information being relayed to the farmers. Additionally, the influence of lighting and background effects should not be underestimated as they can also contribute to the dissemination of false information. In one of the associated approaches each animal was equipped with a GPS collar to enable the transmission of their location and condition to the drone's receiver. Additionally, the minimum average distance between the drones and animals can be calculated by utilizing the density-based clustering algorithm DBSCAN and K-Means clustering algorithm [40].

9.5 CONCLUSION

The use of drones in agriculture offers potential solutions to various challenges faced by farmers, such as labor unavailability, extreme weather events, inadequate fertilizer application, and crop diseases. Drones can be used for irrigation, crop monitoring, soil and field analysis, and bird control, providing valuable data for more accurate problem identification and solution creation. Drones have been increasingly used for monitoring and assessing crops, orchards, and forests, especially for managing stresses like water, diseases, nutrition deficiencies, and pests. The information captured during drone flights can be extracted and analyzed to improve agricultural practices. Drones are advantageous for pesticide application, as they allow for precise crop dusting, reducing costs and minimizing pesticide exposure to manual laborers. This method also has the potential to improve pest management in both small and large-scale crop fields through accurate site-specific applications. Mapping and soil analysis can be conducted manually or automatically using drones, providing valuable information for further analysis of soil conditions and crop yield estimation. Unmanned Aerial Vehicles (UAVs), including multi-rotor drones, are capable of autonomous flight and are operated through radio communication, making them suitable for various agricultural applications. Though ground scouting can be difficult and agricultural areas are frequently large, utilizing drones in agriculture is very beneficial. This chapter gives a thorough summary of the major developments in this area, focusing especially on the techniques used to extract data from the photos taken by drones.

The prevalence of several plant diseases is a significant hindrance to crop quality and growth rate. The lack of precise information in the input samples makes the detection and classification of agricultural leaf diseases a challenging and time-consuming operation. The process of classifying crops is further complicated by variations in the size, location, and structure of the affected crop section, as well as by noise and blurriness in the input photos. A strong strategy based on deep learning and drone technology is suggested to deal with these issues. One technology that shows a lot of promise for increasing agricultural cost-efficiency is automated plant diagnosis. Drone effectiveness is limited by a number of factors, including the trade-off between speed and resolution and an inadequate amount of annotated training data. Plants are identified and categorized as possibly unhealthy or healthy using a machine learning identification. After that, the potentially ill plant photos are given actual diagnoses using a machine learning classifier. UAVs offer a convenient solution in areas where equipment and labor are difficult to access.

REFERENCES

1. Yallappa, D., Veerangouda, M., Maski, D., Palled, V., and Bheemanna, M. (2017) Development and Evaluation of Drone Mounted Sprayer for Pesticide Applications to Crops, in 2017 IEEE Global Humanitarian Technol. Conf., San Jose, CA, USA, 1–7.
2. Qin, W., Xue, X., Zhang, S., Gu, W., and Wang, B. (2018) Droplet Deposition and Efficiency of Fungicides Sprayed with Small UAV against Wheat Powdery Mildew, J. Agric. Biol. Eng., vol. 11, no. 2: 27–32.
3. Spoorthi, S., Shadaksharappa, B., Suraj, S., and Manasa, V. K. (2017) Freyr Drone: Pesticide/Fertilizers Spraying Drone-An Agricultural Approach, 252–255. 10.1109/ICCCT2.2017.7972289
4. Susitra, D., Jebaseeli, E. A. E., Chitturi, V. K., and Chadalavada, V. (2020) Design and Development of an Hexacopter for Fertilizer Spraying in Agriculture Fields. J. Phys.: Conf. Ser. 1706. 012053. 10.1088/1742-6596/1706/1/012053
5. Xue, X., Lan, Y., Sun, Z., Chang, C., and Hoffmann, W. C. (2016) Develop an Unmanned Aerial Vehicle Based Automatic Aerial Spraying System, Comput. Electron. Agric., vol. 128, 58–66.
6. Yanliang, Z., Qi, L., and Wei, Z. (2017) Design and Test of a SixRotor Unmanned Aerial Vehicle (UAV) Electrostatic Spraying System for Crop Protection, Int. J. Agric. Biol. Eng., vol. 10, no. 6: 68–76.
7. Tang, Y., Hou, C. J., Luo, S. M., Lin, J. T., Yang, Z., and Huang, W. F. (2018) Effects of Operation Height and Tree Shape on Droplet Deposition in Citrus Trees Using an Unmanned Aerial Vehicle, Comput. Electron. Agric., vol. 148, 1–7.
8. Jinru, Xue and Su, Baofeng (2017) Significant Remote Sensing Vegetation Indices: A Review of Developments and Applications, Journal of Sensors, vol. 2017. 1–17. 10.1155/2017/1353691
9. Wilson, N. R., Norman, L. M., Villarreal, M., Gass, L., Tiller. R., and Salywon, A. (2016) Comparison of Remote Sensing Indices for Monitoring of Desert Cienegas, Arid L. Res. Manag., vol. 30, no. 4: 460–478.
10. Patel, P. N., Patel, M .A., Faldu, R. M., and Dave, Y. R. (2013) Quadcopter for Agricultural Surveillance, Adv. Electron. Electr. Eng., vol. 3, no. 4: 427–432.
11. Duan, T., Chapman, S. C., Guo, Y., and Zheng, B. (2017) Dynamic Monitoring of NDVI in Wheat Agronomy and Breeding Trials Using an Unmanned Aerial Vehicle, F. Crop. Res., vol. 210, 71–80.
12. Kerkech, M., Hafiane, A., and Canals, R. (2018) Deep Leaning Approach with Colorimetric Spaces and Vegetation Indices for Vine Diseases Detection in UAV Images, Comput. Electron. Agric., vol. 155, 237–243.
13. Bendig, Juliana, Bolten, Andreas, and Barteh, Georg. (2012) Introducing a Low-Cost Mini-UAV for Thermal and Multispectral Imaging, in Int. Arch. Photogramm. Remote Sens. Spatial Inf. in. Sci., XXXIX-B1. 345-349. 10.5194/isprsarchives-XXXIX-B1-345-2012.
14. Wu, K., Rodriguez, G. A., Zajc, M., Jacquemin, E., Clement, M., de Coster, A., and Lambot, S. (2019) A New Drone-Borne GPR for Soil Moisture Mapping, Remote Sens. Environ., vol. 235, 111456.
15. Huuskonen, J. and Oksanen, T. (2018) Soil Sampling with Drones and Augmented Reality in Precision Agriculture, Comput. Electron. Agric., vol. 154, 25–35.
16. von Hebel, Chris, Reynaert, Sophie, Pauly, Klaas, Janssens, Pieter, Piccard, Isabelle, Vanderborght, Jan, Kruk, Jan, Vereecken, Harry, and Garre, Sarah. (2021) Toward High-Resolution Agronomic Soil Information and Management Zones Delineated

by Ground-Based Electromagnetic Induction and Aerial Drone Data, Vadose Zone Journal, 20. 10.1002/vzj2.20099.

17. Pandjaitan, N. H., Rau, M. I., Sutoyo, K. T. U., Waspodo, R. S. B., and Mustikasari, N. (2019) Efficient Mapping for Fodder Cultivation with Commercial Small Drone, in Int. Conf. Digital Agric. Land Consum., Bogor, Indonesia, vol. 335, 1–8.
18. Christiansen, M. P., Laursen, M. S., Jørgensen, R. N., Skovsen, S., and Gislum, R. (2017) Designing and Testing a UAV Mapping System for Agricultural Field Surveying, Sensors, vol. 17, no. 12: 1–19.
19. Chen, P. C., Chiang, Y. C., and Weng, P. Y. (2020) Imaging Using Unmanned Aerial Vehicles for Agriculture Land Use Classification, Agric., vol. 10, no. 9: 1–14.
20. Al-Thani, N., Albuainain, A., Alnaimi, F., and Zorba, N. (2020) Drones for Sheep Livestock Monitoring, 672–676. 10.1109/MELECON48756.2020.9140588
21. Li, X. and Xing, L. (2019) Reactive Deployment of Autonomous Drones for Livestock Monitoring Based on Density-based Clustering, 2421–2426. 10.1109/ROBIO49542.2019.8961763
22. Rivas, A., Chamoso, P., González-Briones, A., and Corchado, J. M. (2018) Detection of Cattle Using Drones and Convolutional Neural Networks, Sensors, vol. 18, no. 7: 1–15.
23. Xu, B., Wang, W., Falzon, G., Kwan, P., Guo, L., Sun, Z., and Li, C. (2020) Livestock Classification and Counting in Quadcopter Aerial Images Using Mask R-CNN, Int. J. Remote Sens., vol. 41, no. 21: 8121–8142.
24. Nar, P., Amin, S. S., Banerjee, S., Garg, V., and Pardasani, A. (2020) Autonomous Multifunctional Quadcopter for Real-Time Object Tracking and Seed Bombing in a Dynamic Environment, in Emerging Technologies for Agriculture and Environment, Subramanian B., Chen S S. and Reddy K R. (Eds.). Singapore: Springer, 199–211.
25. Andrio, A. (2019) Development of UAV Technology in Seed Dropping for Aerial Revegetation Practices in Indonesia. 1OP Conference Series: Earth and Environmental Science, 308. 012051. 10.1088/1755-1315/308/1/012051.
26. Manoharan, A. K., Telagam, N., Mohankumar, N., Ismail, K. M., and Rajasekar, T. (2020) Design and Implementation of Real-Time Amphibious Unmanned Aerial Vehicle System for Sowing Seed Balls in the Agriculture Field, Int. J. Emerg. Technol., vol. 11, no. 2: 213–218.
27. Qing, T, Ruirui, Z., Liping, C., Min, X., Tongchuan, Y., and Bin, Z. (2017) Droplets Movement and Deposition of an Eight- Rotor Agricultural UAV in Downwash Flow Field, Int. J. Agric. Biol. Eng., vol. 10, no. 3: 47–56.
28. Xinyu, X., Kang, T., Weicai, Q., Yubin, L., and Huihui, Z. (2014) Drift and Deposition of Ultra-Low Altitude and Low Volume Application in Paddy Field, Int. J. Agric. Biol. Eng., vol. 7, no. 4: 23–2
29. Shaikh, M. S., Ali, S. I., Deshmukh, A. R., Chandankhede, P. H., Titarmare, A. S., & Nagrale, N. K. (2024). AI Business Boost Approach for Small Business and Shopkeepers: Advanced Approach for Business. In S. Ponnusamy, M. Assaf, J. Antari, S. Singh, & S. Kalyanaraman (Eds.), Digital Twin Technology and AI Implementations in Future-Focused Businesses (pp. 27–48). IGI Global, USA. https://doi.org/10.4018/979-8-3693-1818-8.ch003
30. Ali, S. I., Kale, G. P., Shaikh, M. S., Ponnusamy, S., & Chouhan, P. S. (2024). AI Applications and Digital Twin Technology Have the Ability to Completely Transform the Future. In S. Ponnusamy, M. Assaf, J. Antari, S. Singh, & S. Kalyanaraman (Eds.), Harnessing AI and Digital Twin Technologies in Businesses (pp. 26–39). IGI Global, USA. https://doi.org/10.4018/979-8-3693-3234-4.ch003

31. Shaikh, M. S., Chandrawat, U. B., Choudhary, S. M., Ali, S. I., Ponnusamy, S., Khan, R. A., & Sheikh, A. G. (2024). Harnessing Logistic Industries and Warehouses with Autonomous Carebot for Security and Protection: A Smart Protection Approach. In S. Ponnusamy, M. Assaf, J. Antari, S. Singh, & S. Kalyanaraman (Eds.), Harnessing AI and Digital Twin Technologies in Businesses (pp. 239–257). IGI Global, USA. https://doi.org/10.4018/979-8-3693-3234-4.ch017
32. Ali, S. I., Shaikh, M. S., Ponnusamy, S., & Chouhan, P. S. (2024). Technological Collaboration, Challenges, and Unrestricted Research in the Digital Twin: Digital Twin Technology. In S. Ponnusamy, M. Assaf, J. Antari, S. Singh, & S. Kalyanaraman (Eds.), Harnessing AI and Digital Twin Technologies in Businesses (pp. 380–399). IGI Global, USA. https://doi.org/10.4018/979-8-3693-3234-4.ch028
33. Ali, S. I., Shaikh, M. S., Shahane, S., Sharma, K., & Macwan, K. (2025). The Era of Metaverse and Generative Artificial Intelligence. In L. Gaur (Ed.), Responsible Implementations of Generative AI for Multidisciplinary Use (pp. 29–44). IGI Global, USA. https://doi.org/10.4018/979-8-3693-9173-0.ch002
34. Ali, S. I. & Shaikh, M. S. (2025). The Ethical Dilemma of Using (Generative) AI to Science and Research. In L. Gaur (Ed.), Responsible Implementations of Generative AI for Multidisciplinary Use (pp. 249–264). IGI Global, USA. https://doi.org/10.4018/979-8-3693-9173-0.ch009
35. Shaikh, M. S., Ponnusamy, S., Ali, S. I., Wanjari, M., Mungale, S. G., Ali, A., & Baig, I. (2024). AI-Based Advanced Surveillance Approach for Women's Safety. In S. Ponnusamy, V. Bora, P. Daigavane, & S. Wazalwar (Eds.), Wearable Devices, Surveillance Systems, and AI for Women's Wellbeing (pp. 13–25). IGI Global, USA. https://doi.org/10.4018/979-8-3693-3406-5.ch002Mungale
36. Mungale, S. G., Mungale, N. G., Shaikh, M. S., Mungale, M. S., Wazalwar, S. S., Wanjari, M. M., & Jichkar, R. A. (2024). Safeguard Wrist: Empowering Women's Safety. In S. Ponnusamy, V. Bora, P. Daigavane, & S. Wazalwar (Eds.), Wearable Devices, Surveillance Systems, and AI for Women's Wellbeing (pp. 192–205). IGI Global Scientific Publishing, USA. https://doi.org/10.4018/979-8-3693-3406-5.ch012
37. Sheikh, M. S., Ali, A., Baig, I., Bhanodia, P. K., et al. (2024). Harnessing Logistic Industries Using Autonomous Carebot for Smart Surveillence, Protection and Security. In: Al-Turjman, F. (Ed.), The Smart IoT Blueprint: Engineering a Connected Future. AIoTSS 2024. Advances in Science, Technology & Innovation. Springer, Cham. https://doi.org/10.1007/978-3-031-63103-0_20
38. Kitey, H., Chandankhde, P., Jajulwar, K., Shahnawaz Shaikh, M. and Fatinge, P. M. (2024). Solar Power Generation Technique and its Challenges – A Comprehensive Review, Grenze International Journal of Engineering and Technology, 10(1), 2778–2786. Grenze ID: 01.GIJET.10.1.122, Grenze Scientific Society, 2024
39. Chopkar, P., Wanjari, M., Jumle, P., Chandankhede, P., Mungale, S. and Shaikh, M. S. (2024). A Comprehensive Review on Cotton Leaf Disease Detection using Machine Learning Method, Grenze International Journal of Engineering and Technology, 10(2), 239–245. Grenze ID: 01.GIJET.10.2.537, Grenze Scientific Society, 2024
40. Shaikh, S., Choudhry, A., Wadhwani, R. (2016). Analysis of Digital Image Filters in Frequency Domain. International Journal of Computer Applications. 140, 6 (April 2016), 12–19. DOI=10.5120/ijca2016909330

10 Deployment of 5G Drones

Technical Challenges and Solutions for Special Uses Cases for Smart Agriculture

G. Ranjitham and Nagarajan Sathish Kumar

10.1 INTRODUCTION

Due to technological developments there is a paradigm shift in agriculture technology – the Unmanned Aerial vehicle (UAV) or 'drone', which is an electronic controlled device and capable of flying with the assist of autopilot and the co-ordinates of GPS. Drones can be controlled using RF signals using a small unit called Remote. Table 10.1 provides the summary of the development cycle of agriculture.

There are more cutting-edge technologies that are applied to increase the productivity or yield with less time. Aerial data is one of the important technologies for the changes taking place. New drone-based agriculture policy decisions have resulted in good faming production. 5G technology provides faster download (Gbps), high bandwidth (1000 MHz) and good latency (1mSec). [8, 14] Figure 10.2 shows the main features of the 5G network.

10.2 AGRICULTURAL DRONES

As per the report by Research and Markets, it is expected that the usage of drones by Indian farmers will rise at a CAGR of 30% during the year 2025. [9] The following are the main factors in preferring drone technologies for smart farming. To cite a few, low hovering ability, low cost, light weight, planning and control. Due to advancement in navigation like LIDAR and high resolution camera mounting capabilities, good resolution of multispectral and hyper spectral images of good resolution images are acquired with low complexity. It is proven to be an effective tool for better efficiency, profitability and protecting environment.

10.2.1 Drone Technology

Drone technology depends on the amalgamation of software and hardware to collect, process, and analyze agricultural data[1]. Prepared with special sensors and cameras, 5G drones hover over farmland, catching high-resolution pictures and

DOI: 10.1201/9781003487005-10

TABLE 10.1
Summary of Agriculture Development Cycle

Sl.No	Agriculture Evolution	Year	Technology Evolution
1.	Agriculture 1.0	1974	Use of human and animal resources, organic farming.
2.	Agriculture 2.0	1950	Use of powered machinery, pesticides and fertilizers.
3.	Agriculture 3.0	1992	Application of high speed and high yield machines, swoil analysis, weather forecast and analysis.
4.	Agriculture 4.0	2020	Use of Internet of things, Big Data, AI, Robotics.
5.	Agriculture 5.0	2024	Features application of nano science, bio technology, ML, DL,Data Analytics, NLP, drones.

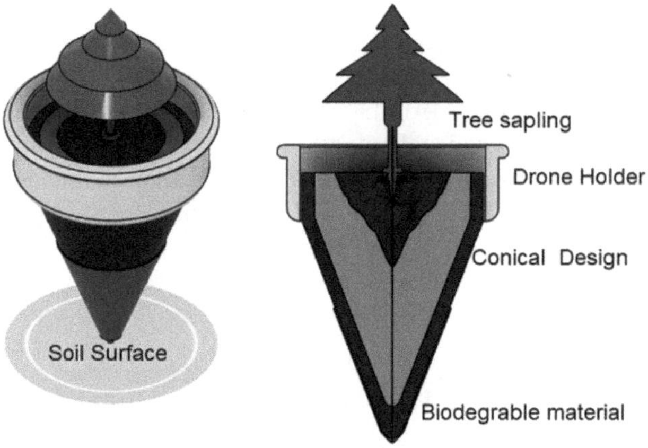

FIGURE 10.1 Forces associated with drone movements.

collecting various data points. These data are then processed using refined algorithms to produce actionable visions for farmers. The main components of drones include battery, flight controller, frame, sensor, motors and propellers. The design architecture of a drone for agricultural applications includes many factors like the aerodynamic shape of propellers, weight of total drone unit, strength, electric motor, electric speed controller, RF transmitter and receiver, and software interface for data analysis. Figure 10.1 categorises the forces acting on a drone.

Fluid dynamics decides the forces acting on drones. The thrust generated by each propeller is perpendicular to the plane of rotation of propellers. Due to its large mass the total drone weight will act towards the direction of Earth's gravitational force.

The higher the mass of the drone, the higher the power requirements to lift and move it. The vertical force acting on the drone is called lift. This is required to move and lift the drone against Earth's gravity. All the propellers generate the required force to kick start the 5G drone from the Earth's surface. Force acting on the drone opposite to the direction of the drone's motion is called drag. Drag is due to the viscosity of air and pressure difference. Based on propeller architecture, drones are categorized as follows: Bicopter (2 propellers), Triple copter (3 propellers), Quad copter (4 propellers), Hexa Copters (6 propellers) and Octa copter (8 propellers). A Quad copter is the more renowned option for agricultural applications.

10.2.2 Categories of UAV

The classification is based on the architecture, range and altitude, power and speed. Table 10.2 summarizes the various types of UAVs and their features and applications.

10.3 REQUIREMENTS

This section will illustrate the various technical and non-technical requirement for the deployment of 5G drones in smart farming applications.

10.3.1 Regulations

Regulations/Laws for the operation of 5G drones must be set out for the full pledged operation and integration of 5G drones in smart agriculture. The laws are framed by the international Civil and Aviation, European Tele -Communication union, and other foreign bodies. Drone usage land laws shall be published, and able to reach the farmers to get the benefits of the technology. In India, DGCA, a statutory body, regulates civil aviation. Farmers can acquire drone certification.

10.3.2 5G as a Seamless Communication Enabler

Sensors in farms links to drones through a gateway before recording cloud and data visualizations [2]. They also enable a large number of connected devices through IoT. Some of the performance factors for 5G drones in agriculture are latency, energy consumption, reliability of link, radio link failure, energy minimization and speed. 5G drones also transfer data wirelessly during GPS denied locations or Non Line of Sight conditions (N-LOS), Blue tooth Low energy (BLE), Software Defined Radio (SDR) and networks. 5G networks will improve the connectivity, routing and other performances.

10.3.3 Data

5G drones for various agricultural purposes capture high definition multispectral and hyper-spectral images through the high resolution mounted cameras, extract files and

TABLE 10.2
Features and Application of UAVs

Sl. No	Classification parameter	Label Features	Features	Applications
1.	Size and weight	Nano	0-250 Gms.	Aerial imaging and crop and soil monitoring.
		Micro	250 Gms-2 Kgs	Surveillance and combat attack
		Medium	2Kgs-25Kgs	Pesticide spray
		Large	25Kgs-200Kgs	Goods delivery
2.	Altitude and ange	Hand held	600m and 2kms	Agriculture applications
		Close	<1500m - <10Kms.	Videography, agriculture applications
		NATO	<3000m - <50Kms	Entertainments
		Tactical	<5500m - <160Kms	Firefighting and patrolling
		MALE (Medium altitudes long endurance)	<9150m - <200Kms	Videography, disaster managements.
		HALE (High altitudes long endurance)	>9150m - Indefinite range	Aerial mapping and long distance monitoring
		Hypersonic	15200m - >200Kms	Large coverage, Infrastructure surveillance, Goods delivery
3.	Wings and Rotors	Single rotor	Long time and hovering, Large coverage, vertical take off	Large-scale mapping
		Fixed wing	No hovering	Large-scale serial mapping
		Multi-rotor	Short endurance time, Small payloads,	Aerial mapping and long-distance monitoring
		Hybrid VTOL	Long endurance	Aerial imaging and Crop and soil monitoring.

analyze data. All such operations require a large amount of data storage which is to be stored in the drone unit. Generally a 2Gb–8Gb data storage unit is required with data security. Hence technologies like cloud storage, virtual storage and software defined storage will be an alternative solutions.

10.3.4 Data Security

Data breach can affect farmers. The major requirement is to protect the data from cyber-attacks [5,6]. There are some possibilities of attacks namely Denial of Service, Eavesdropping, ransomware attacks, phishing attacks and malwares which can corrupt the software and make the 5G drone malfunction. Block chain algorithms [4] are applied to make the data more secure. John Deere (2017), JFC international (2021), Loaves and Fishes (2020), Home Chef (2020), Fonterra (2020), Australian wine (2020), Solar Winds (2020) are some the attacks in food and agriculture industries. Usage of strong passwords (Special Characters) for log-in applications, double authentications, a Licensed Software, Antivirus software usage of Linux based Operating system can be used for security. It is desirable to take advance measures towards the risk of cyber-attacks and take up the security measures.

10.3.5 Flight Time and Energy Consumption

The average flight duration of UAV is about 15–25 minutes with the assist of a sole battery. The UAV with many rotors are battery powered. Thus the flight time is depended on aerodynamic design, wind speed, weight, avionics, UAV speed, direction of motion, air density, payload, actuators used. The energy requirement of the UAV can be determined by mathematical and empirical models. At equilibrium the drone hovering is given in equation (1).

Hovering thrust (h_T) = Drone Weight (W_D)

$$\text{Drone Weight } (W_D) = mg \tag{1}$$

Where, m= mass in Kg, g= acceleration due to gravity constant 9.8.

It is inferred that the greater the weight, the greater will be the UAV thrust requirements. As payload increases the flight time decreases. Battery capacity and flight time are related using equation (2).

$$\text{Flight time } (T_1) = \text{Battery capacity (Wh) / Power (W)} \tag{2}$$

Increasing the efficiency of the propeller system will decrease the power drawn from the battery and thus improve the flight time [3].

10.3.6 Technology Popularization

The application of the drone will be made available to farmers under technology popularization schemes. Farmers can be educated about using drones for acreage coverage. Aerial image is one of the common application of drones.

10.4 TECHNICAL CHALLENGES AND SOLUTIONS

10.4.1 Challenges

This section presents the challenges and solutions for the impact of transmission conditions when the radio channel properties vary continuously. In certain conditions drones act as a relay element. The following are the some of the factors affecting transmission capacities.

10.4.1.1 Antenna Orientation

Antenna orientation in drones can affect the transmission capabilities. For successful actions of drone operation the signal strength is an important factor. To achieve this, antenna orientation must support the best signal quality. Remote control (or) two monopole antennas are used from drone remote control. It is fine to adjust the orientation of the antenna as per the position of the antenna and to maintain strong wireless link. Drones prefer circularly polarized antennas. The general frequencies preferred for drones are in the 2.4GHz, 5.8 GHz frequency band. Electro-Magnetic Interference (EMI) will affect the signal quality. Figure 10.2 shows the Antenna orientation for the remotely connected drones.

10.4.1.2 Intermodulation Distortion

Intermodulation Distortion (IMD) occurs when two or more radio frequencies combine to form new third frequency defined as undesired frequency. Figure 10.3 shows the concept of intermodulation distortion. This is an unwanted frequency production or sideband leading to adjacent channel interference. The IMD frequency band can be determined using equation (3). When radio waves are tangled, the third-order intermodulation distortions are important because they fall in the area of the unique frequency of the radio wave. Figure 10.7 shows the existence of IMD for two given frequency F_1 and F_2.

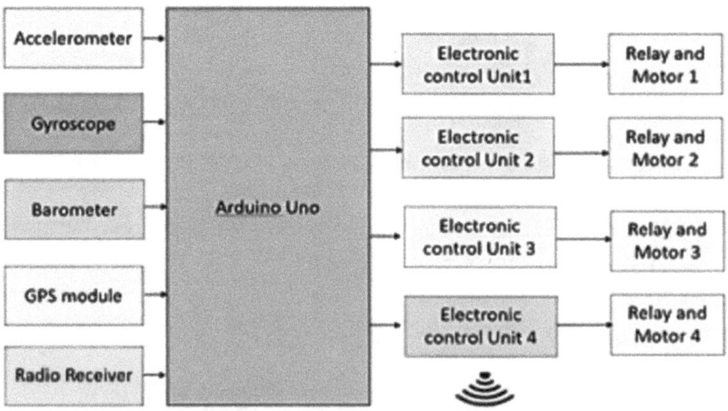

FIGURE 10.2 Antenna orientation for drones.

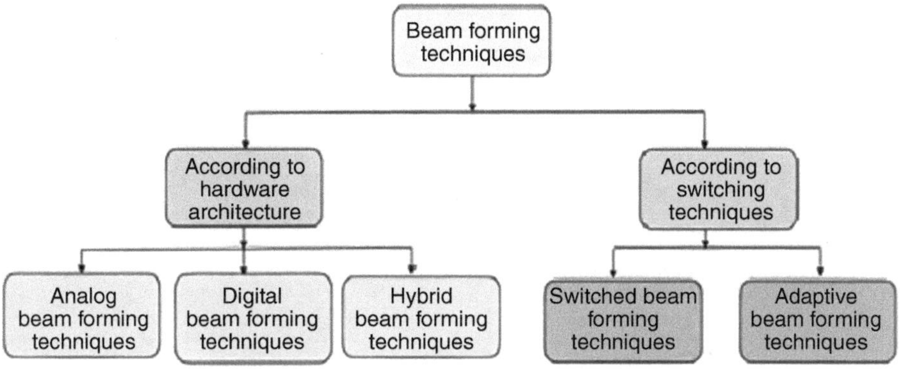

FIGURE 10.3 Intermodulation distortions (frequency vs power characteristics).

TABLE 10.3
Frequency Bands for Drone Communication

Name of the Drone Band	Frequency range & bandwidth	Power limit
M-Bands	1625 MHz - 2510 MHz 1.625 GHz - 2.51 GHz	≤ 10 m Watt
Short range devices	433 MHzs and 1.74 MHz	≤ 10 m Watt
Intermediate short range	915 MHzs and 26.0 MHzs	≤ 100 m Watt
ISM Band	2.4 GHz	≤ 100 m Watt
ISM Band	5.7 G Hz	≤ 100 m Watt

$$IMD = (2F_1 - F_2) \text{ and } (2F_2 - F_1) \quad (3)$$

where, F_1 and F_2 are the two given frequencies.

10.4.1.3 Spectrum Availability

Table 10.3 defines the band and frequency range and the power limits of drones used for agriculture applications. Remote control drones use uplink and downlink frequencies. They often use 900 MHzs band. This band is also preferred for armature radio and cannot create any interference.

10.4.1.4 Data Traffic Congestion

Figure 10.4 shows the Internet of Drone Architecture (IoD). Multi drone communications lead to network congestion [11,15]. Many drones try to share packet information via nodes to the cloud. The architecture consists of three layers, i.e. Drone layer, Fog and Cloud layer respectively. IoD is a communication protocol which can dynamically control air space and provide navigation services. Transfer of a continuous packet will lead to congestion. Intensive use of IoD results in massive computation and communication, resulting in network congestion. Drones need to

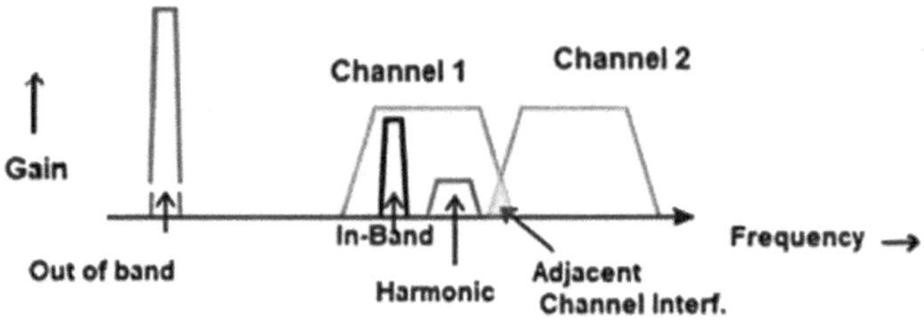

FIGURE 10.4 IoD architecture.

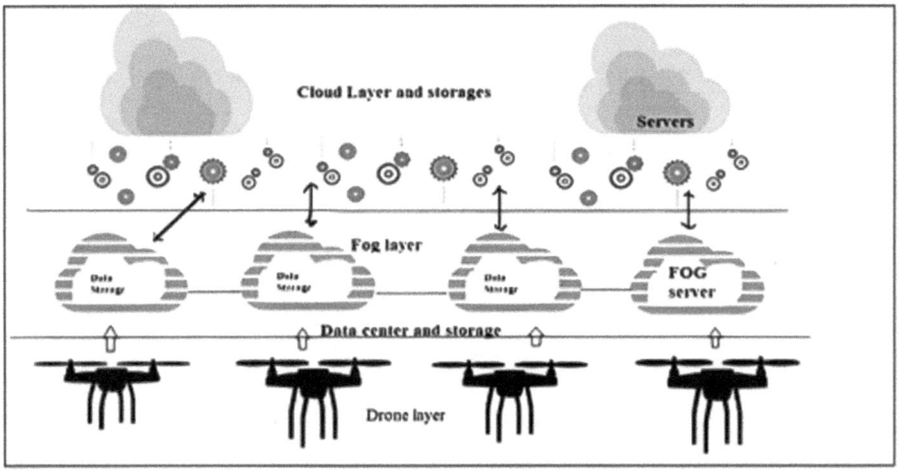

FIGURE 10.5 Types of interference in drone environment.

update the serves about the locations and emergency services, which is a challenging issue in IoD environment. Reliable communication and timely packet delivery are essential in these conditions

10.4.1.5 Multipath Fading

This is a condition when a receiver tracks a variety of versions of the same signal with variations in signal amplitude, phase and angle of arrival. This type of channel impairment caused is called multipath fading. An example for this is an echo. Due to the atmospheric conditions and obstacles, the transmitted signal undergoes reflection or refraction and reaches the receiver with multiple versions of same signal.

10.4.1.6 Interference

The interference in wireless is broadly classified in six classes as follows. In-band (IBI), Out-band (OBI), Adjacent Channel (ACI), Co-channel (CCI), downlink and uplink interference. If the signal strength of the interference signal is greater than desired signal, then the desired signal is degraded. [13] Figure 10.5 models all the type of interference in a drone environment. CCI is a type of interference caused when more devices are operated in same frequency channel. It can occur in voice channel and control channels. There are two main reasons for this type of interference: poor weather circumstance and inappropriate frequency planning. Such interference can be reduced by a frequency reuse technique and proper frequency planning. ACI is caused due to signals from other bands mixing with the desired signal. There are two significant causes of ACI: use of the same or similar frequency and unrelated power emission from adjoining channels. Proper filter design, dynamic channel allocation and provision of guard bands can minimize such types of interference. OBI arises in the channel due to poor filter design causing imperfect filtering, due to non-linearity in operation and leakage of signal power. This type of design also produces harmonics in other bands. Good filter designs can eliminate such problems. In-band interference relates to the frequency overlap of band pass filter characteristics. Downlink and uplink interference are related to satellite communication, to the signal from Earth station to satellite and from satellite to Earth. Transmission of data from a cell tower to the user's equipment or a terminal relates to uplink and data transfer from a user equipment (drone) to ground station is called downlink.

10.4.1.7 Weather Conditions

There are many weather conditions namely rain, fog and snow conditions which will affect wireless transmission and reception of signal. There is a large amount of transmitter power requirement based on climatic conditions, number of base stations and distance between base stations. Radio link failure is possible due to the weather conditions. This further introduces channel varying properties, link drops, addition of noise, and signal interference.

10.4.2 Solutions

5G drones provide scalable and flexible wireless coverage in the field. There are many possible and feasible solution to tackle the challenges. Let us discuss each one of them. To overcome the problem of channel variations 5G drones can use several techniques as discussed below.

10.4.2.1 Hybrid Beam Forming

Beam forming, also known as spatial filtering, in which antenna arrays are used to direct or focus the signal energy in a particular direction or a receiver for faster and reliable wireless communication. This method offers many advantages such as concentrating a beam in a particular direction, high power towards the receiver, fast information exchange, no transmission errors and minimum signal interference. Figure 10.6 shows the broad categories of beam forming techniques.

In analog beam forming, a group of feed lines form a phased array through a single input data. A purposeful phase shift is applied at the input and reaches the antenna. This method is applicable for low cost and low beam. More transmit antennas give more gain and provide a narrow beam. This method has only a set of Data converters is available. The digital beam forming method has individual data converters and a

FIGURE 10.6 Types of beam forming in 5G drones.

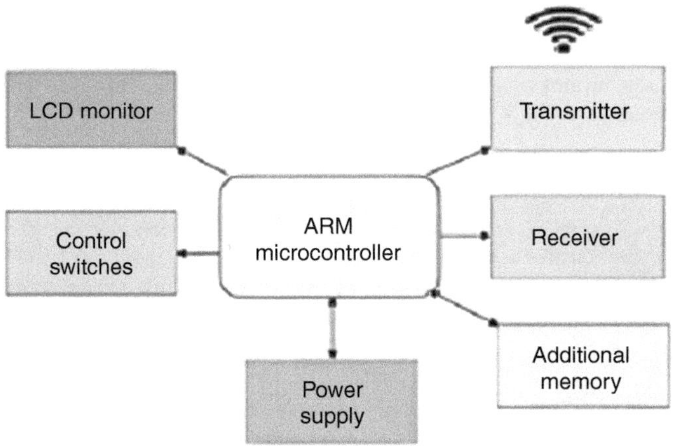

FIGURE 10.7 Concept of beam forming.

transceiver. Hence many signals are generated and fed to the antenna, producing a multiple beam. In this method it is possible to transmit data concurrently to many receivers. This scheme essentially needs more hardware and signal processing units and it is expensive. Hybrid beam forming is an effective approach with less design complexity, less energy consumption, and a concentrated radiation focus. It unites the advantages of analog beam forming and digital beam forming approaches. The antenna array has sub-arrays of analog beam forming elements, and digital merging elements of the sub-array signals. Figure 10.7 demonstrates the concept of beam forming.

10.4.2.2 Network Slicing

Network slicing in 5G permits the construction of multiple distinctive virtualized and logical networks over a shared platform. It allows dynamic resource allocation in the 5G network. A single slice can have its individual logical topology, security features and performance factors. Different slices can be developed and devoted to diversified purposes. The data path and resources are allocated statically. Three classes are identified for the network slicing: Enhanced Mobile broad band (eMBB) which produce the most traffic in a mobile environment, Massive Machine type communication (mMTC) which connects billion and billions of devices to the core network, and Ultra reliable low-latency communication (urLLC) which provides low latency in V2x communication. The 5G network provides many important functions such as authorization, service registration and de-registration, inter service, recovery and selection through suitable communication interfaces reinforced with 5G QoS structure. Some Quality indicators are default window averaging, maximum burst volume of data, packet error, packet delay, priority level and bit rate.

10.4.2.3 Massive MIMO (Multiple Input and Multiple Output) mMIMO

A Massive MIMO is a wireless technology that has many transmitting antennas and receiving antennas to enhance performance factors like capacity of channel, reduction of interference, throughput, quality, energy efficiency and spectral efficiency for mobility environments. The term 'massive' indicates the number of antennas. The base stations having large antenna arrays function independently. This provides a large number of users concurrently in operation. The main objective of mMIMO is to create many spatial signal paths. Device to device communication is also possible.

10.4.2.4 Dynamic Spectrum Access (DSA)

DSA is a real-time method of sharing many frequency spectrums without interference, having the re-configurability of usage of spectrum holes making a transmission possible without extra use of bandwidth. 5G drones can uses DSA to choose the frequency band and maintain the transmission during the change in channel environment.

10.4.2.5 Artificial Intelligence (AI) and Machine Learning (ML)

AI plays an important role in agriculture. AI Algorithms analyze the issues and provide possible solutions to farmers. Data produced are stored and processed by AI algorithms for monitoring, control and to predict the solution. Currently in Market AI and ML based multifunction agribots, customized agribots, and smart tractors are available [10,12]. AI provides novel solutions for complex problems. Machine learning algorithms are applied to increase the accuracy and efficiency of decision making in agriculture systems and thus reduce the risk and cost.

10.4.2.6 Switching in Hybrid Wireless Networks (HWN)

It is a type of network switches, router hubs, gateways, and other server-based networks. 5G drones can switch from one wireless network to another. It is an integration of Mobile Adhoc Network and infrastructure wireless network. Data packet routing methods can improve scalability and capacity. Multi-hop technology is also used in HWN.5G drones, which can switch from one wireless network to another with variations in network like Wi-Fi, satellite network, VANET, and optical wireless network. Such networks offer more advantages in terms of scalability, flexibility, trouble shooting and integration to virtual cloud.

10.4.2.7 Antenna Diversity

Antenna diversity employs many antennas to enhance the reliability of wireless links to counterattack multipath effects at the receiver side. This has an advantage that while an antenna is in deep fade situation the other antenna has a high signal strength. 5G drones can employ multiple antennas to establish a link with reasonable to high channel capacity under the variation of channel conditions. In space diversity, duplicate versions are transmitted through many antennas. Space diversity evades shadowing and fading. Signal scanning for high strength is done at receiver with a threshold of 'n' number of signals. Such a scheme is called feedback diversity. A similar method where the weights are made to unify and obtain the output is termed Equal Gain combining. Macroscopic diversity enables multicast services. The diversity scheme also helps to fight shadowing (Slow Fading). Different polarizations used between a pair of transmitters and receivers producing two diversity. This is termed as polarization diversity. Pattern diversity is also known as angular diversity. The time diversity scheme is the re-transmission of original signals at discrete intervals of time. There will be no rise in transmission power. It will be ensured that the coherence time is greater than re-transmission time. The frequency diversity method is derived by sending the information with multiple carrier signal frequency. More receivers are required hence it is a high cost system. It has a large bandwidth requirements. Figure 10.8 shows the classification of antenna diversity.

10.4.2.8 Cloud-Based Radio Access Network (C-RAN)

It is a virtualized-based cloud radio access network that works on cloud servers. 5G drones combine to share the cloud resources as per the channel changing conditions. It improves the performance of networks. The core unit is connected to central pool. The central pool has any Baseband Processing (BP) units which are connected to the radio head through front haul links. The front haul links are low latency, high capacity links. C-RAN architecture relies on virtualization technologies to consolidate BBUs, enabling resource sharing and dynamic allocation among multiple cell sites.

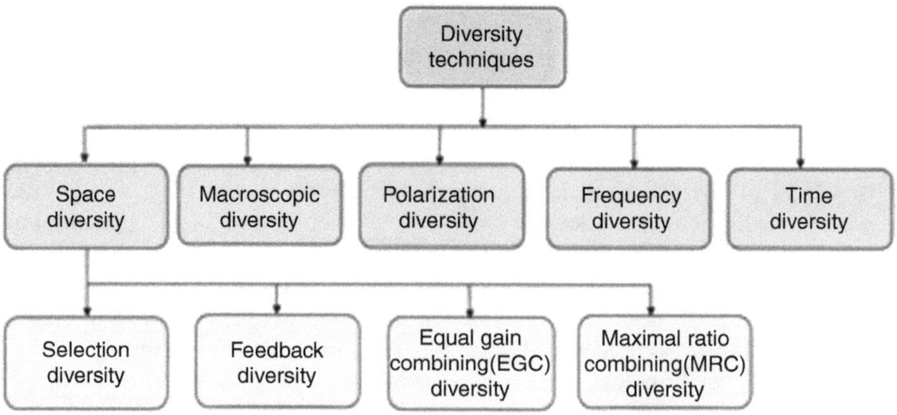

FIGURE 10.8 Classification of diversity schemes.

10.4.2.9 Multi-Hop Communication

A unique way of data transmission by using a chain of intermediate hops or nodes to spread the communication range. 5G drones can combine to relay data over a hopping system. It increases the signal strength by marinating the transmitting capabilities during the variations in channel conditions. There are two classes, namely Single Hop Network (SHN) and Multi-Hop Network (MHN). Less energy consumption, low latency, and simplicity are a few advantages of SHN but it has some limitations such as range and scalability. MHN delivers long range, is scalable and more reliable but there is more energy consumption and it introduce ore latency in communication. Depending on the distance, power, applications, scalability, either SHN or MHN is deployed.

10.4.2.10 Software-Defined Radio (SDR)

This is a technology for implementing radios' working functions through a variety of hardware and software on programmable devices which include Field programming Gate Array (FPGA), Digital Signal Processors (DSP), Programmable system on Chip. SDR used software to build radio hardware and to achieve multiple functions, namely channel equalization, modulation and frequency hopping. Using SDR, 5G drones can adjust to channel conditions.

10.5 5G DRONES IN SMART FARMING

This section presents some of the application of 5G drones in smart farming.

10.5.1 Soil and Field Analysis

5G Drones are applied in smart farming to examine the soil fertility, landscape conditions, and moisture level in soil of the agricultural land. This further helps the farmer to plan for seed sowing, to prepare an irrigation chart, and manage the fertilizers. A knowledge of blends of soil balance is essential for the farmer in terms of fertility and nutrients. 5G drones provide the soil analysis and provides information about the moisture and nutrient levels. A plant needs macro and micro nutrients like carbon, hydrogen, oxygen, potassium, boron, nitrogen, phosphorus with pH conditions. This is called digital soil mapping. It helps farmers to gather information about soil classification and its fertility. The remote sensing cameras mounted on drones capture the multispectral information and create the multispectral soil map. This allows the farmers to take decisions about the crop plantation and conduct land management strategies.

10.5.2 Bombing for Planting Trees and Sowing Seeds

5G drones regularize the planting sequence as per programmed dimensions. They cover a large area at faster speed and save the labor cost for the famer which can be deployed in challenging landscape areas. They can save tractor fuel and also reduce pollution. 5G drones can be used to plant trees or crops in remote areas by

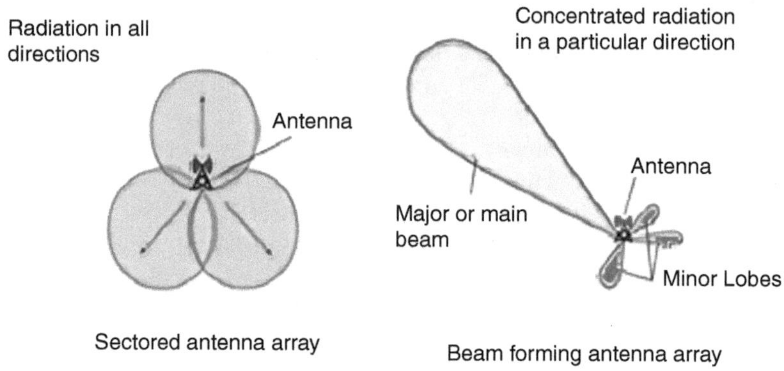

FIGURE 10.9 Seedling planted through bombing approach.

throwing biodegradable seed pods or seed bombs. Thus, they can be used for the restoration of degraded lands by planting trees, and also for reforestation as well as afforestation activities. 5G drones use a seed bombing method for planting trees and sowing seeds. LIDAR technology is applied for mapping the seed zone and spreading seed according to the seed area ratio. The drone launches a shaped cone made of biodegradable material with a tree sapling into the soil. This is as shown in Figure 10.9.

10.5.3 Crop Monitoring Using RF Control

5G drones are used to monitor the crop growth, insects and weeds. It is an effective and cost lucrative method that captures multispectral images. Deep Learning (DL) algorithms are effectively applied for weed detection, crop pest, and prediction of leaf diseases. DL algorithms require large memory, high computation with additional peripheral resources [9]. 5G drones help the farmers to monitor the crops continuously throughout the season. Crop irrigation is important for effective yield productivity. Accurate and timely data will help the farmer to take the control of irrigation using soil moister. Digital cameras help in capturing the land wetness and dryness condition apart from moister sensors. Arduino uno controls the relay and motor circuit depending on soil moister. Figure 10.10 (a–b) has two parts namely drone hardware and remote unit. The hardware system consists of a camera, sensors, microcontroller, solar panel, Actuators, RF module, and transmitter. Other sensors include a barometer, accelerometer and camera. The MEMS module is used to monitor the acceleration, velocity, orientation and displacement of the drone. The barometer provides a provides a digital 24-Bit data for pressure and temperature; the RF module is used to offer wireless communication between drone and external unit. Actuator, propellers and a brushless motor help the drone to propel. Arduino Uno is the main controller responsible for signal processing from sensors and remotely controls the entire system. This controller receives data and decodes commands and initiates action through control signals to and from the drone.

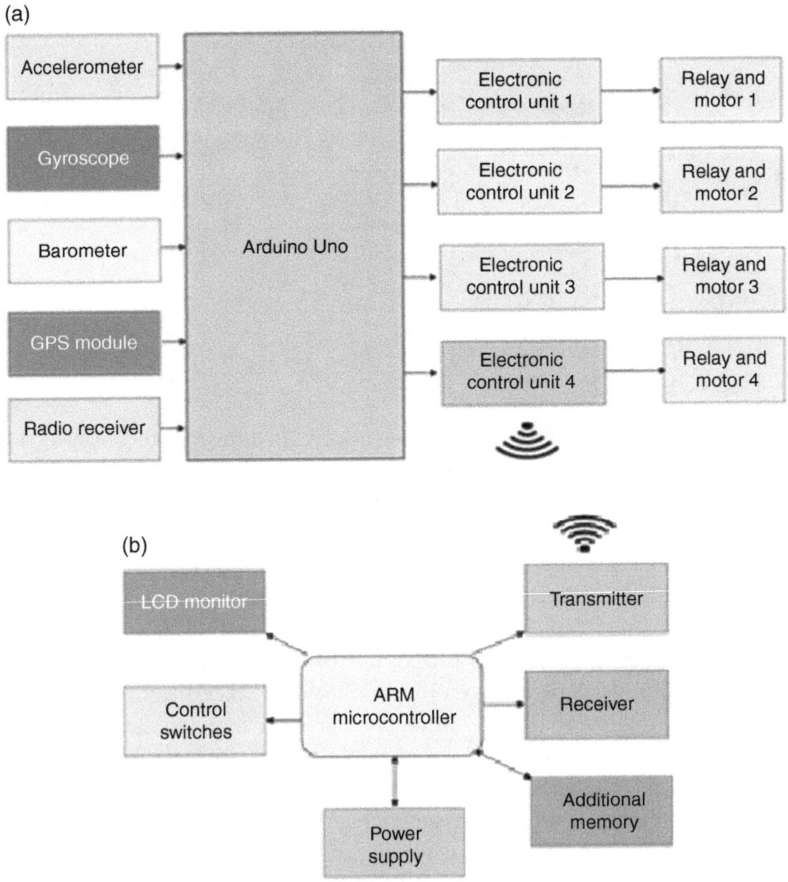

FIGURE 10.10 (a-b) Drone hardware and remote unit.

10.5.4 Crop Spraying

5G Drones are used for pesticide and fertilizer spraying. Figure 10.11 shows a drone spraying pesticides at low altitude flight mode. The parameters are drone velocity, flight duration, battery backup, pesticide volume, spray height and width, nozzle droplet size and pressure, wind speed and directions. The amount of chemicals to be sprayed in the field is adjusted based on the crop condition or the degree of diseases. Thus drones provide a solution for precision farming. The chemical dissemination is done at a faster pace in the field than the manual method, and saves on the cost of laborers for days. The accuracy of the flight height and velocity can be controlled by the trained operator and remote control. The spray can be controlled by the type of nozzle. Nozzle size can be easy controlled according to the pesticide or fertilizer. The larger the height, the larger the field area of coverage.

FIGURE 10.11 Crop spraying drone.

10.6 CONCLUSION

This paper demonstrates the potential uses of 5G drones in precision agriculture. The main advantages of using 5G drones are mobility in rainy conditions and capacity to capture multispectral and hyper-spectral images from different altitudes. It is also possible to monitor the crops attacks by pests, weeds and animals, leaves at the micro level in identifying leaf diseases, lack of nutrients in fields and crops. This chapter has presented the drone structure, design of drones and the data recording and retrieving mechanisms in an IoD environment. Future scope will be based on battery and RF control to increase the flight time, use of Blue tooth, Low energy Schemes to save power, high resolution IR, fuel technology, advanced remote sensing, interactive processing, auto image annotation, power chip management, wireless charging, higher drone endurance, aerial supervision, 6G and beyond for communication to achieve faster and higher data rates, night vision cameras and usage of LIDAR. The chapter also demonstrated the issues, challenges and key technologies and all possible solutions for smart farming using the 5G Drones. The paper concludes with some potential application like digital soil mapping, crop field monitoring, planting of tress and seedlings using bombing method, crop spraying and fertilizing,crop monitoring and other applications.

REFERENCES

1. Adaka S, Mandala N, Maitya PP, and Mukhopadhyaya A (2022) "Drones: A modern breakthrough for smart farming," *Food Science Report*, vol. 3, no. 9, pp. 54–57.
2. Ahirwar S, Swamkar R, Bhukya S, and Namwade G (2019) "Application of DRONE in agriculture," *International Journal of Microbiology and Application Science*, vol. 8, no. 01, pp. 2500–2505.
3. Galkin B, Kibilda J, and DaSilva LA (2019) "UAVs as mobile infrastructure: Addressing battery Lifetime," *IEEE Communications Magazine*, vol. 57, no. 6. pp. 132–137.
4. Hakak S, Khan WZ, Gilkar GA, Imran M, and Guizani N (2020) "Securing smart cities through block chain technology: Architecture, requirements, and challenges," *IEEE Network*, vol. 34, no. 1, pp. 8–14.
5. Homtz 8 M. (2020) "Cyber attack-resilient cyber physical systems," *IEEE Transactions on Security & Privacy,* vol. 18, no. 1. pp. 55–60.
6. Xiang JLJ, Jin Y, Jining Yan RL, and Vilang L (2021) "Boost precision agriculture with unmanned aerial vehicle remote sensing and edge intelligence: A survey," *Remote Sensing*, vol. 13, no. 1, pp. 01–31.

7. Kumar et al. (2020) "CIG based stress identification method for maize crop using UAV based remote sensing," *IEEE Sensors Applications Symposium (SAS)* Kuala Lumpur, Malaysia, pp. 836–846.
8. Kim J, Kim S, Ju C, and Son HI (2019) "Unmanned aerial vehicles in agriculture: A review of perspective of platform, control, and applications," *IEEE Access*, vol. 7, pp. 105100–105115.
9. Malik AW, Rahman AU, Qayyum T, and Ravana SD (2020) "Leveraging fog computing for sustainable smart farming using distributed simulation," *IEEE Internet of Things Journal*, vol. 7, no. 4, pp. 3300–3300.
10. Motlagh NH, Taleb T, and Arouk O (2016) "Low-altitude unmanned aerial vehicles based internet of things services: Comprehensive survey and future perspectives," IEEE *Internet of Things Journal,* vol. 3 no. 6, pp. 899–922.
11. Kal C, Zhou H, Yi Y, and Huang W (2020) "Collaborative cloud-edge-end task offloading in mobile-edge computing networks with limited communication capability," *IEEE Transaction on Cognitive and Communication Network,* vol. 7, pp. 624–634.
12. Perumal R and Nagarajan SK (2022) "A machine learning-based compressive spectrum sensing in 5G networks using cognitive radio networks," *International Journal of Communication Systems*, vol. 35, no. 16, pp. e5302.
13. Kiruhika V, Kumar S and Vimal SP (2022) "Performance Analysis of Interference Cancellation and Traffic Scheduling Algorithm (ICTSA) with Mobility Management in 5G Emergency Networks Proceedings," 2nd *International Conference on Smart Technologies*, Communication and Robotics 2022, STCR 2022.
14. Ramakrishnan P, Sivagurunathan PT, and Sathishkumar N (2021) "A comprehensive survey on effective spectrum sensing in 5G wireless networks through cognitive radio Networks," *Journal of Physics*. Conference Series Open Access, vol. 17, no. 19, Article number 012070. doi 10 1088/1742-6596/1717/1/012070.
15. Vimal SP, Sathish Kumar N, Kasiselvanathan M and Gurumoorthy KB (2021) "Smart irrigation system in agriculture," *Proceedings Journal of Physics Conference Series*, vol. 1917, no. 1.

11 Applications of AI-Enabled Drones in Agriculture

Veningston K, Selvan C, Ronalda M, and Sangeetha S K B

11.1 INTRODUCTION

The integration of AI with drone technology [1] marks a significant evolution in the agricultural sector, offering a promising avenue to address the challenges of modern farming. As global populations rise and land suitable for growing crops diminishes, the pressure to increase agricultural productivity while conserving resources has never been more significant. AI-enabled drones emerge as pivotal tools in this context, harnessing the power of advanced data analytics to optimize every aspect of farm management, i.e. from seeding and irrigation to pest control and harvest timing.

AI-enabled drones are not merely remote-controlled aerial vehicles but highly sophisticated data collection platforms equipped with sensors, cameras, and Global Positioning System (GPS) technologies. Drones capture high-resolution aerial images and multispectral data that provides detailed insights into crop conditions, pest control, soil health, and environmental factors. The real strength of AI lies in its ability to process and analyze this vast amount of data rapidly, offering actionable insights beyond the capability of human analysis. For instance, through pattern recognition algorithms, AI can identify areas of a field suffering from nutrient deficiencies or water stress and suggest precise interventions. This technology also extends to the precise application of agricultural inputs. By analyzing data collected in real-time, drones can execute targeted spraying of water, fertilizers, or pesticides, reducing the overall use of these inputs while maximizing their effectiveness. This study not only saves cost but also mitigates the environmental impact of farming, aligning with sustainable agriculture principles. Moreover, AI-driven drones play a critical role in planting and seeding. They can perform these tasks with high precision [2] and at a fraction of the time and cost required by traditional methods, particularly in terrains that are challenging to access. This capability is vital in regions prone to adverse weather conditions, where timely planting is crucial for crop success. The surveillance capabilities of AI-enabled drones extend beyond crops to livestock management. Drones monitor animal health and behavior, warning early about potential disease outbreaks or breeding problems. This surveillance helps manage large herds more efficiently, ensuring optimal health and productivity. Figure 11.1 depicts the capabilities of AI-enabled drones in analyzing and monitoring plant health in real time [3].

DOI: 10.1201/9781003487005-11

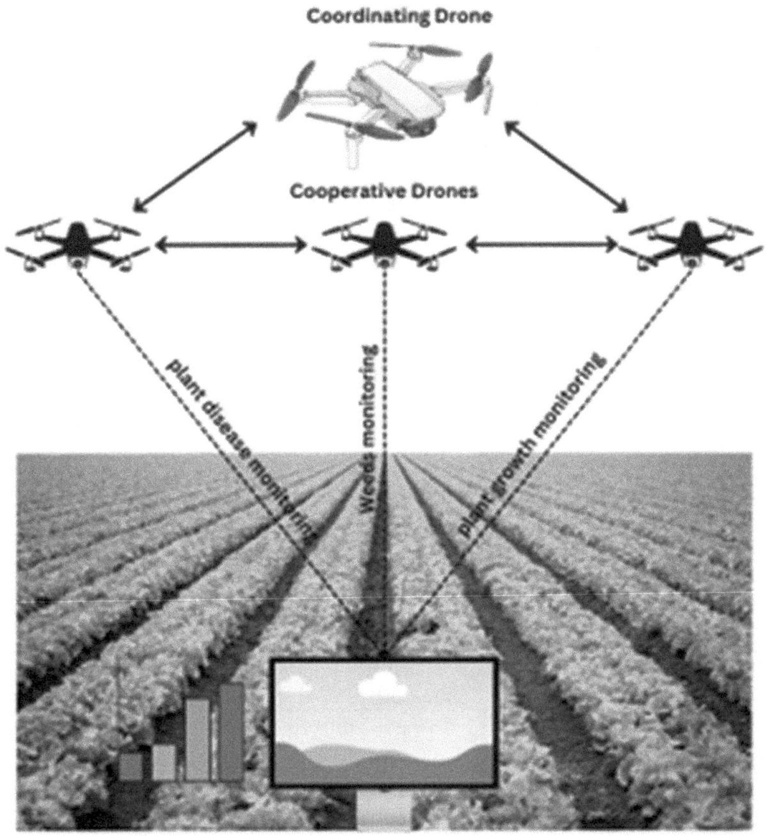

FIGURE 11.1 AI-enabled drones analyzing and monitoring plant health in the agriculture field.

The data integration feature of AI-enabled drones represents a quantum leap in agricultural management. AI algorithms can predict future farm conditions by synthesizing drone-collected data with other data sources like weather forecasts and soil sensors. This predictive power allows farmers to make proactive decisions, enhancing crop yields and reducing the risk of crop failure. Therefore, promising AI-enabled drones [4] transform agriculture into a more efficient, sustainable, and productive industry. This investigation empowers farmers with real-time insights from the field and paves the way for a future where farm management is highly data-driven and precise, minimizing waste and maximizing yield.

The introduction discusses integrating AI with drone technology in modern farming. The remaining chapter content is organized as follows: Section 11.2 describes the AI technologies relevant to agriculture and drone components, including sensors, cameras, and GPS; Section 11.3 describes how drones are used for monitoring crop health and pest control; Section 11.4 describes the role of drones in agricultural resource management; Section 11.5 describes the technologies used for drone-assisted pest, disease controls and seeding; Section 11.6 details how drone data is integrated with other agricultural data sources; and Section 11.7 summarizes the chapter with the role of AI-enabled drones in advancing sustainable agriculture.

11.2 KEY TECHNOLOGIES AND COMPONENTS

In AI-enabled drones for agricultural use, several critical technologies and components interplay to optimize their functionality and effectiveness, as shown in Figure 11.2.

11.2.1 Description of AI Technologies Relevant to Agriculture

AI technologies have revolutionized agricultural practices by enabling drones to analyze complex datasets and identify hidden patterns. The following subsections delve into specific AI methodologies and their agricultural applications.

11.2.1.1 Computer Vision

Computer Vision (CV) allows drones to interpret and analyze visual information from their surroundings. In agriculture, CV is used to assess crop health, detect weeds, and monitor pest invasions, all through image analysis. This is done by analyzing changes in color, shape, size, and other hidden features compared to healthy norms.

11.2.1.2 Data Analytics

AI excels in extracting insights from complex datasets through Exploratory Data Analytics (EDA). For agricultural drones, analytics can provide recommendations on crop rotation, optimal planting times, and resource allocation based on historical data and real-time inputs. Machine Learning (ML) further empowers drones to process and analyze the vast amounts of data they collect, enabling them to recognize patterns, predict outcomes, and make data-driven decisions [5]. For example, ML models can differentiate between healthy and unhealthy plants in real-time and identify field areas requiring more water.

In AI-enabled drones for agricultural use, several critical technologies and components interplay to optimize their functionality and effectiveness, as shown in Figure 11.2, which highlights these essential elements in detail.

11.2.2 Overview of Drone Components

Drones used in agriculture typically consist of several key components: a lightweight, durable frame, high-resolution cameras or multispectral sensors for capturing data, GPS for navigation, rotors for flight stability and control, and a battery for power. Advanced models may also include communication technologies for real-time data transmission.

11.2.2.1 Multispectral Cameras

Multispectral cameras capture data beyond what is visible to the human eye, including infrared and ultraviolet light, providing detailed insights into plant health and water usage, as depicted in Figure 11.3, which illustrates the functionality and applications of these sensors in agricultural drones.

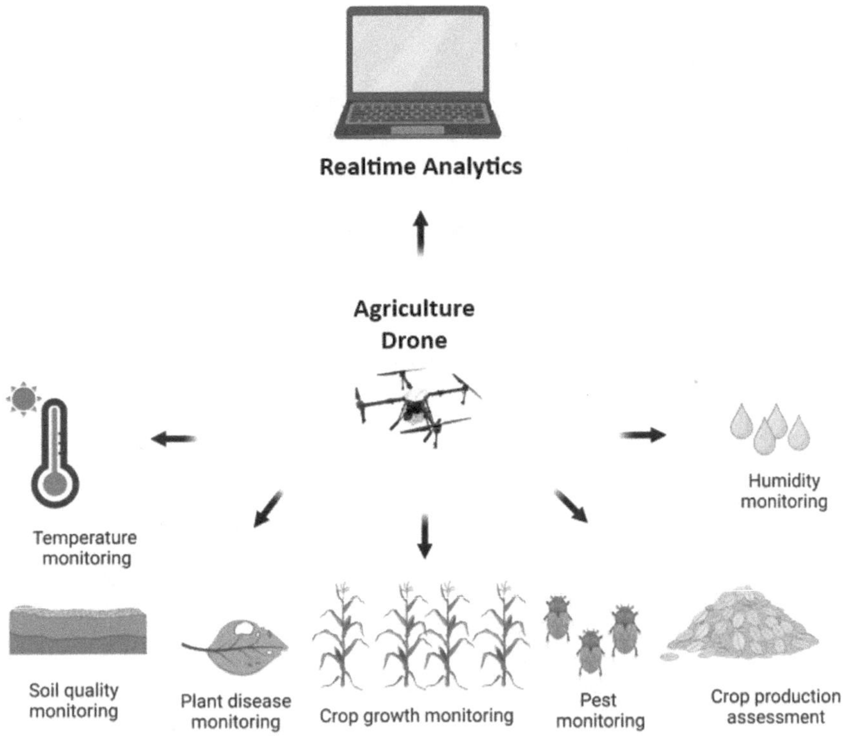

FIGURE 11.2 Critical components in the AI-enabled agriculture drone.

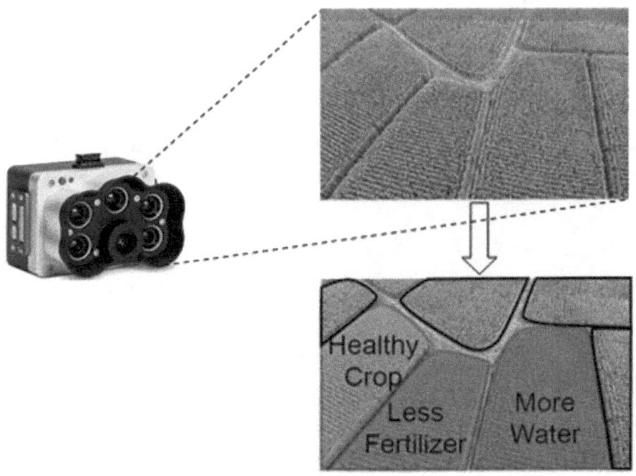

FIGURE 11.3 Multispectral camera sensors.

FIGURE 11.4 Thermal imaging sensors detect temperature variation across a field.

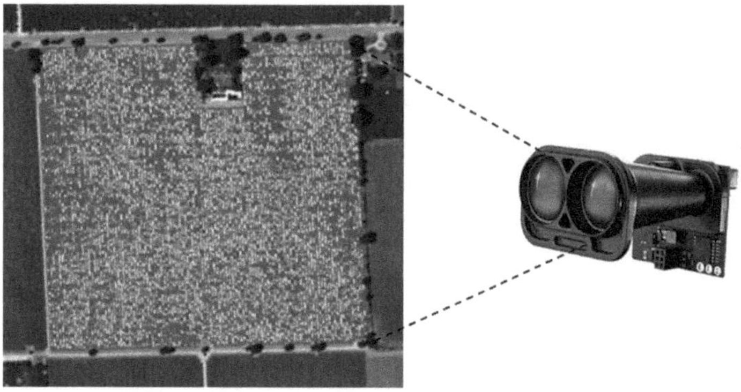

FIGURE 11.5 LiDAR used for agriculture.

11.2.2.2 Thermal Imaging Sensors

Thermal imaging sensors in agricultural drones detect temperature variations across fields to identify irrigation needs, monitor plant health, and manage pests. Figure 11.4 illustrates how this technology enhances crop management by optimizing water usage and detecting early signs of disease progression. Thermal sensors can even increase animal presence in crop fields.

11.2.2.3 LiDAR Sensors

Light Detection and Ranging (LiDAR) is a robust remote sensing [6] tool that provides high-resolution, three-dimensional (3D), and spatial information about agricultural land. These sensors help create detailed 3D terrain maps, which help analyze crop structure, plant biomass, and soil properties. LiDAR technology offers farmers innovative methods for gathering and examining data on their farmland's soil type and composition. This crucial information assists farmers in assessing which crops are most suitable for cultivation on their land and deciding the appropriate types of fertilizer.

LiDAR technology provides high-resolution, three-dimensional spatial information about agricultural land that enables detailed analysis of crop structure, plant biomass, and soil properties. Figure 11.5 illustrates how LiDAR is utilized to gather and examine critical data for optimizing agricultural practices.

11.2.2.4 Global Positioning System (GPS)

GPS navigation is essential for the precise operation of drones; GPS enables path planning and accurate positioning, which is crucial for tasks like seeding, spraying, or detailed field mapping.

11.2.2.5 Real-Time Kinematic (RTK) Navigation

This advanced GPS technology provides centimetre-level accuracy, enhancing the precision of operations like planting seeds or applying nutrients. RTK is a GPS correction technique that offers real-time corrections to location data while the drone captures imagery.

11.2.2.6 Wireless Transmission

Drones communicate wirelessly to send collected data to farm operators or integrated systems, using technologies like Wi-Fi, Bluetooth, or even advanced cellular networks like 4G/5G.

Global Positioning System (GPS) and Real-Time Kinematic (RTK) navigation play a crucial role in ensuring the precise operation of drones to enable tasks such as seeding, spraying, and detailed field mapping. Figure 11.6 highlights the integration of GPS and RTK technologies in agricultural drones for enhanced accuracy and efficiency.

11.2.2.7 Cloud Computing

Cloud computing services enable the storage and processing of huge amounts of aerial image data collected by drones that facilitate access to this data for real-time analysis and long-term scalable monitoring.

FIGURE 11.6 Use of GPS and RTK in Agriculture drone.

11.2.2.8 Data Processing Applications

Data processing applications take raw aerial images captured by drones and convert them into actionable insights, often presented through user-friendly interfaces that allow farmers to make prompt decisions.

11.2.3 How do AI and Drone Technology Synergize?

AI and drone technology synergize in agriculture, creating a powerful tool for enhancing farm management and operations. Drones with sensors and cameras capture vast amounts of aerial image data from the fields, including images and environmental conditions. AI processes [7, 8, 9] this data, utilizing algorithms to analyze and interpret complex datasets quickly and accurately. AI enhances the precision of drones by enabling them to execute complex tasks autonomously. For example, AI algorithms can determine optimal flight paths for drones to ensure complete field coverage for tasks like spraying, seeding, or monitoring. AI allows for real-time data processing directly on the drone or immediately after the data is sent to the farmer's computer or cloud system. This capability enables immediate decision-making, such as adjusting crop management practices on the fly based on current field conditions. By integrating AI with drones, predictive models can be developed based on historical data and real-time inputs. These models forecast future agricultural outcomes, such as the best times for planting or the potential spread of pests and diseases. AI algorithms are particularly effective in enhancing the capabilities of imaging technologies used in drones, such as multispectral and thermal imaging. These algorithms interpret the aerial images to assess crop health, pest control, detect weeds, and monitor water stress levels with high accuracy.

AI helps integrate the data collected by drones with other data sources, such as soil sensors, aerial images, and weather forecasts. This integration provides a holistic interpretation of the farm's status, facilitating more informed and strategic farm management decisions.

11.3 CROP MONITORING

Crop monitoring towards precision agriculture involves utilizing drones equipped with standard and multispectral cameras, to gather aerial data on crop health and soil conditions. This data-driven approach enables farmers to precisely manage inputs like water, fertilizers, and pesticides, targeting specific areas of need. By analyzing this data, farmers can detect early signs of plant disease, plant nutrient deficiencies, and water stress, allowing timely interventions to optimize crop yields and reduce waste. Integrating these technologies fosters a more efficient, sustainable approach to farming, thus significantly improving resource management and the overall productivity of agricultural operations.

11.3.1 Evolution of Drones in Agriculture

The evolution of drones in agriculture can be traced through several critical phases, each marked by technological advancements and expanding applications. The evolution of drones in agriculture is comprehensively summarized in Figure 11.7.

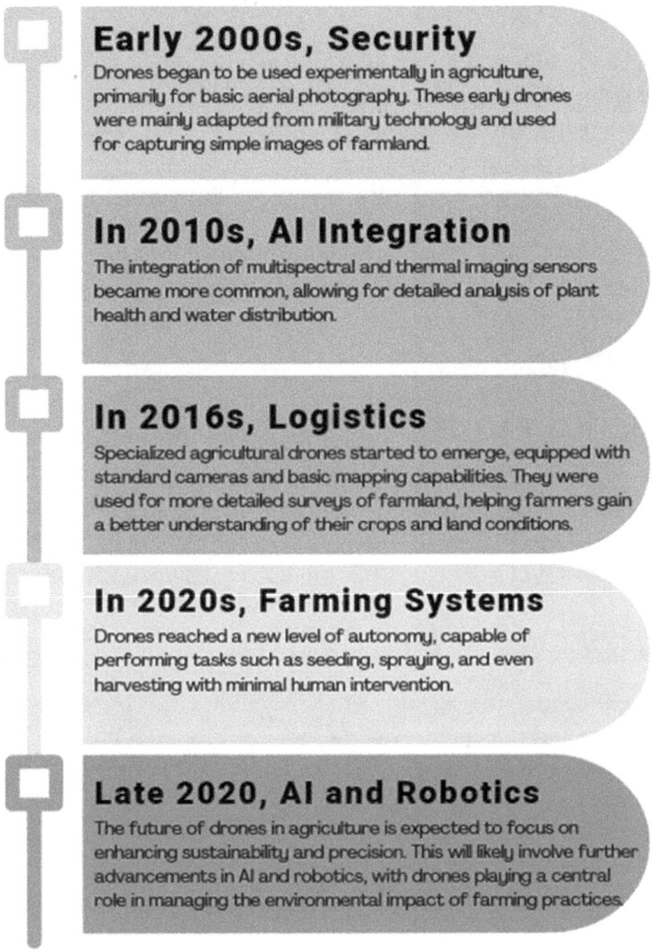

FIGURE 11.7 Evolution of drones.

11.3.2 Drones for Monitoring Crop Health

Drones for monitoring crop health have become a crucial component in the toolkit of modern farmers, providing them with detailed insights that can significantly enhance their agricultural practices. Data from drones can be integrated with other information sources, such as soil health sensors and weather data, within a farm management system. This holistic view enables predictive analytics, helping farmers anticipate issues before they arise and plan for better crop management strategies [10].

The evolution of drones in agriculture has progressed through several critical phases, each marked by technological advancements and expanding applications. Figure 11.7 provides a comprehensive summary of these developments to exhibit the transformative applications of drones in modern farming.

11.3.3 Impact on Crop Yield

Drones significantly enhance crop yields by enabling precise agricultural practices. They provide detailed insights into crop health, allowing for targeted interventions such as optimized irrigation, accurate pesticide application, and effective nutrient management. This precision reduces resource wastage and boosts plant health, directly increasing crop productivity and sustainability, leading to higher yields and more efficient farming operations.

Advancements in AI and drone technology continue to improve the efficiency of crop health monitoring and control. Future trends include using drones in autonomous swarms for large-scale monitoring and integrating AI to provide even more refined analytics.

11.4 AGRICULTURAL RESOURCE MANAGEMENT

Agricultural resource management leverages technology to optimize water, fertilizers, and pesticides, enhancing sustainability. By integrating data from various modes, including drones and sensors, farmers can tailor their practices to the specific needs of their crops, reducing waste and improving yields while minimizing environmental impact.

11.4.1 Drones in Detecting Irrigation Needs

Drones with thermal imaging sensors detect irrigation needs by identifying field temperature variations and indicating water stress levels among crops. This technology allows farmers to precisely target irrigation efforts, ensuring water is applied efficiently where needed, reducing wastage, and promoting optimal plant growth. This precise irrigation management leads to better resource use and healthier crops.

11.4.2 Efficient Water Usage Enabled by AI Analysis

AI analysis enhances water usage efficiency by processing data from sensors and drones to pinpoint exact irrigation needs. By predicting moisture levels and detecting areas of water stress, AI enables targeted watering practices, reducing waste and maximizing crop health. This smart irrigation approach ensures that water resources are managed optimally, leading to significant savings and sustainable farming operations [11, 12].

11.5 PEST, DISEASE CONTROL, AND SEEDING

AI-enabled drones are revolutionizing how farmers handle pest and disease control and seeding processes, making these tasks more efficient and effective [13].

11.5.1 Methods for Pest Detection and Disease Prevention

AI-enabled drones regularly surveil crops by capturing high-resolution aerial images at varying altitudes processed using AI algorithms to detect early signals of pest infestations or disease outbreaks. These drones can identify specific problem areas with great accuracy. Once a problem is detected, drones can be deployed to apply pesticides or other treatments precisely where needed, minimizing the amount of chemicals used and reducing the impact on the environment and non-targeted crop areas. This targeted approach conserves resources and helps prevent the spread of pests and diseases.

11.5.2 Technologies Used for Drone-assisted Seeding

Drones are also transforming the seeding process. Equipped with dispensers, these drones can plant seeds at precise locations and optimal soil depths across a field. AI algorithms optimize flight patterns and seeding rates based on the terrain and existing soil conditions, which are analyzed from data collected by the drone's sensors. This method ensures uniform seed distribution and optimal plant density and can be particularly beneficial in rough terrains where traditional planting methods are challenging. Furthermore, drone seeding can drastically reduce the labor required for planting, lowering costs and increasing efficiency.

11.5.3 Advantages of Using Drones for Planting Under Challenging Terrains

Using drones for planting under challenging terrains offers significant advantages. Specially designed drones access steep, uneven, or soft soils where traditional machinery cannot go, minimizing soil compaction and disruption. Drones can quickly and accurately deploy seeds over large areas, increasing planting efficiency and reducing labor costs. This method ensures consistent seed placement and depth, improving germination rates and crop uniformity. Overall, drones enhance the feasibility and productivity of planting in challenging agricultural landscapes.

11.5.4 Effectiveness in Improving Germination Rates

Drones enhance germination rates by ensuring precise seed placement and optimal planting depth tailored to specific soil conditions and seed types. This accuracy improves seed-to-soil contact, which is essential for adequate germination. Additionally, drones ensure proper coverage of agricultural fields that allows effective monitoring of field and ensures seeds will be planted at the ideal time during optimal environmental conditions, further boosting germination success and leading to more uniform crop growth.

11.5.5 AI Capabilities in Identifying and Diagnosing Crop Health Issues

AI capabilities significantly enhance the identification and diagnosis of crop health issues. By analyzing data from drones equipped with advanced imaging technologies, AI algorithms can detect anomalies in plant growth, coloration, and other physiological indicators. These insights allow for the early detection of nutrient deficiencies, water stress, diseases, and pest infestations. AI processes [14] this data quickly, providing farmers with precise, actionable insights that enable targeted interventions. This proactive approach minimizes damage and boosts crop resilience, leading to healthier plants and improved yields. AI's ability to interpret complex datasets makes it indispensable in modern precision agriculture.

11.5.6 Strategies for Targeted Pesticide

Targeted pesticide application strategies using AI and drones focus on precision and efficiency. Drones with sensors detect areas of infestation and stress, allowing for localized pesticide application. This minimizes chemical usage and environmental impact. AI algorithms analyze data to determine optimal spraying times and conditions, enhancing effectiveness. Integrating drone data with weather and soil sensors further refines application accuracy, ensuring pesticides are used only as needed, reducing costs and supporting sustainable farming practices.

11.5.7 Reduction in Labor Costs

Drones significantly reduce labor costs in agriculture by automating tasks [15] like planting, pesticide spraying, and crop monitoring. By handling these tasks efficiently and over large areas quickly, drones minimize the need for manual labor. This automation allows farms to allocate fewer resources to time-consuming activities, leading to substantial savings in labor expenses and enabling workers to focus on more skilled tasks within the operation.

11.6 DATA INTEGRATION AND PREDICTIVE ANALYTICS

Data integration and predictive analytics are transformative agriculture approaches that leverage AI and advanced data management systems [16, 17] to optimize farming practices.

11.6.1 Drone Data Integration with Other Agricultural Data Sources

Data integration in agriculture consolidates information from various sources, including drones [18], soil sensors, and weather stations, into a unified system. This process allows for a holistic view of farm conditions. By integrating various data streams as shown in Figure 11.8 illustrates how these data streams are integrated to provide a comprehensive view of farm conditions, enabling them to make essential, timely decisions [19].

11.6.2 AI-enabled Predictive Analytics

Predictive analytics utilizes the integrated data to forecast farm conditions and yield outcomes. AI models comprehend historical climate and farm data alongside current aerial image data inputs to predict trends and potential issues before they become apparent. Models can forecast crop yields based on weather patterns, crop conditions, and management practices, helping farmers plan storage, labor, and market activities. Predictive analytics can identify the likelihood of pest invasions or disease outbreaks by analyzing patterns and anomalies in the integrated data, allowing preemptive

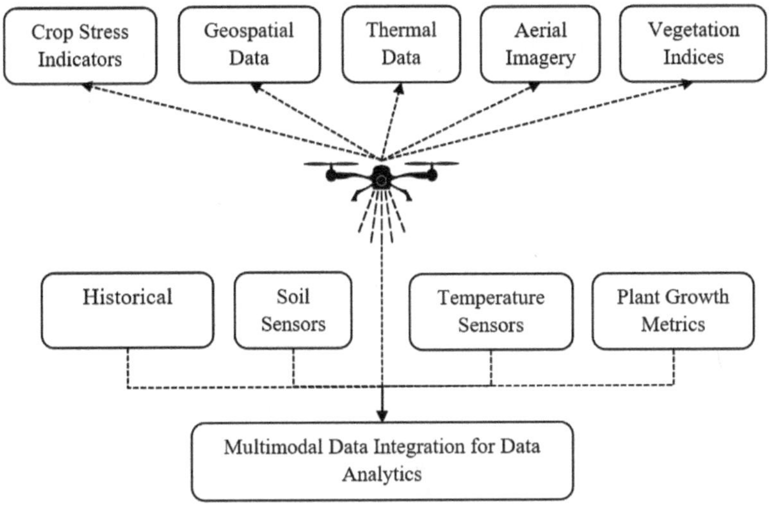

FIGURE 11.8 Drone data integration.

actions. AI can predict the optimal amounts of irrigation, fertilizers, and pesticides to enhance resource utilization.

11.6.3 Potential of AI and Drones in Decision-making Processes

Combining AI and drones revolutionizes agricultural practices by providing real-time data-driven predictive analytics. This integration allows for optimized resource allocation, targeted pest management, and timely interventions, enhancing crop yields and reducing waste. AI-driven insights help farmers anticipate problems, adjust practices, and improve overall farm efficiency, making agriculture more sustainable and economically viable.

11.6.4 Integrative Analytics

Integrative analytics combines data from multiple modalities, such as aerial images from drones and sensor data to comprehensively view farm conditions. This approach provides deeper insights into crop health, pest control, soil status, and environmental factors. By synthesizing this diverse data, integrative analytics aids in making informed decisions, predicting trends, and optimizing resource use. This results in enhanced crop yields, improved efficiency, and reduced environmental impact, making it a pivotal tool in advancing modern precision agriculture and sustainable farming practices.

11.7 CHALLENGES

Initial setup and maintenance costs for drone and AI technology can be prohibitive for small to medium-sized farms, limiting widespread adoption. Managing the vast amounts of data generated and ensuring its security and privacy can be challenging. Farmers and agronomists must navigate data ownership and privacy concerns. The operation of AI-driven drones requires a certain level of technical expertise. Training for farmers and workers is necessary to maximize the benefits of these technologies. Drones face strict regulatory environments that can vary significantly between regions. Compliance with airspace regulations and data usage laws is essential but often complex. Integrating new drone technologies with existing agricultural management systems can be difficult, requiring significant adjustments and compatibility considerations. Technological advancements in battery life and drone durability will enable longer flights and more extensive area coverage, increasing the efficiency of agricultural monitoring and interventions.

11.7.1 Role of AI-enabled Drones in Advancing Sustainable Agriculture

AI-enabled drones are pivotal in sustainable agriculture by optimizing resource usage and enhancing farm management practices.

Drones equipped with AI optimize the application of irrigation, weed control, fertilizers, and pesticides by targeting only the areas that need treatment. This precision reduces the overall consumption of these resources, minimizes overflow into adjacent ecosystems, and lowers the environmental impact of farming. AI-driven drones continuously monitor crop health across vast and varied terrains, detecting issues like water stress, nutrient deficiencies, and disease early. Early detection allows for preventing widespread issues and promoting healthier crops with potentially higher yields using fewer resources. Integrating AI with drones enables collecting and analyzing large datasets that allows farmers to make timely decisions based on real-time data processing. This capability supports planting strategies, harvesting times, and resource management optimized for sustainability goals. AI-enabled drones can help maintain and monitor biodiversity on farmlands. For instance, they can be used for reforestation projects, monitoring wildlife populations, and even managing habitats that support a diverse range of species, contributing to ecological balance.

Drones reduce waste related to the over-application of resources like seeds, water, and chemicals by providing precise information about crop status and needs. This not only saves resources but also prevents unnecessary environmental degradation. Continuous monitoring and data collection help maintain the soil's long-term health. By avoiding over-farming and helping in crop rotation and soil conservation practices, drones contribute to sustainable land management.

11.7.2 Potential Future Developments in Drone Technology and AI in Agriculture

Future developments in drone technology and AI in agriculture are expected to enhance autonomy, increase IoT device integration [2], and improve real-time data processing. Advances may include swarms of drones working collaboratively for large-scale operations, greater precision in plant-specific treatments, and developing AI models that predict agricultural outcomes with higher accuracy. Additionally, improvements in battery technology will likely extend drone flight times, increasing efficiency and coverage. These innovations will continue to drive sustainability and productivity in farming practices.

11.7.3 Benefits of Drones in Agriculture Practices

- Both pest control and seeding are carried out with high precision, reducing waste and increasing the effectiveness of each action.
- AI-driven drones cover agricultural fields and thus save time and reduce the labor traditionally required for planting and monitoring.
- Continuous data collection provides farmers with insights to aid in data-driven analysis about crop management that lead to improved yields.
- Drones help in agricultural practices by reducing the volume of chemicals used for pest control and optimizing water and nutrient use through precise planting.
- AI-enabled drones and robots [20] perform planting, harvesting, weeding, and spraying pesticides without human intervention. These machines operate autonomously, guided by AI algorithms that plan their routes and actions based on real-time data. Ongoing improvements in AI will allow drones to perform more complex tasks with greater accuracy and autonomy, such as advanced pest identification and autonomous robotic harvesting.

11.8 CONCLUSION

As technology advances and becomes more commonplace, the costs associated with drones and AI are expected to decrease, making them accessible to a wide range of applications in agriculture. Further integration of drones with IoT devices and big data analytics will enhance the precision of farming insights, allowing for more dynamic, real-time decision-making across all aspects of farm management. Future developments will increasingly focus on sustainability, using drones to help achieve more environmentally friendly farming practices by minimizing chemical use, optimizing water management, and reducing carbon footprints. As they evolve, drones will further empower farmers to optimize farming operations and promote environmentally friendly practices, significantly impacting global food production.

REFERENCES

1. Su, J., Zhu, X., Li, S., & Chen, W. H. (2023). AI meets UAVs: A survey on AI-empowered UAV perception systems for precision agriculture. *Neurocomputing*, *518*, 242–270.
2. Qazi, S., Khawaja, B. A., & Farooq, Q. U. (2022). IoT-equipped and AI-enabled next generation smart agriculture: A critical review, current challenges and future trends. *IEEE Access*, *10*, 21219–21235.
3. Javaid, M., Haleem, A., Khan, I. H., & Suman, R. (2023). Understanding the potential applications of Artificial Intelligence in the Agriculture Sector. *Advanced Agrochem*, *2*(1), 15–30.

4. Lawrence, I. D., Agnishwar, J., & Vijayakumar, R. (2024). Autonomous Agriculture: The Role of IoT, AI, and ML in Drone Operations. In Rajeev Kumar et al. (Eds.),*Sustainable Development in AI, Blockchain, and E-Governance Applications* (pp. 14–34). IGI Global.
5. Rajak, P., Ganguly, A., Satadal Adhikary, S., & Bhattacharya, S. (2023). Internet of Things and intelligent sensors in agriculture: Scopes and challenges. *Journal of Agriculture and Food Research*, 14.
6. Alkhafaji, M. A., Ramadan, G. M., Jaffer, Z., & Jasim, L. (2024). Revolutionizing Agriculture: The Impact of AI and IoT. In *E3S Web of Conferences* (Vol. 491, p. 01010). EDP Sciences.
7. Kirpach, M., & Riccoboni, A. (2023). AI in agriculture. *Engineering Mathematics and Artificial Intelligence: Foundations, Methods, and Applications*, 331. 1st Edition, Eds. Herb Kunze et al. Chapter 14, pp. 330–364.
8. Alzubi, A. A., & Galyna, K. (2023). Artificial intelligence and the Internet of Things for sustainable farming and smart agriculture. *IEEE Access*, 11, 78686–78692.
9. Shaikh, T. A., Rasool, T., & Lone, F. R. (2022). Towards leveraging the role of machine learning and artificial intelligence in precision agriculture and smart farming. *Computers and Electronics in Agriculture*, *198*, 107119.
10. Gaur, A. S., Raghuvanshi, C. S., & Sharan, H. O. (2024). Smart Prediction Farming Using Deep Learning and AI Techniques. In Rajeev Kumar et al. (Eds.), *Sustainable Development in AI, Blockchain, and E-Governance Applications.* (pp. 152–170). IGI Global.
11. Alazzai, W. K., Obaid, M. K., Abood, B. S. Z., & Jasim, L. (2024). Smart Agriculture Solutions: Harnessing AI and IoT for Crop Management. In *E3S Web of Conferences* (Vol. 477, p. 00057). EDP Sciences.
12. Khan, A., Malebary, S. J., Dang, L. M., Binzagr, F., Song, H. K., & Moon, H. (2024). AI-enabled crop management framework for pest detection using visual sensor data. *Plants*, *13*(5), 653.
13. Martos, V., Ahmad, A., Cartujo, P., & Ordoñez, J. (2021). Ensuring agricultural sustainability through remote sensing in the era of agriculture 5.0. *Applied Sciences*, *11*(13), 5911.
14. Arya, L., & Rastogi, R. (2024). Study on Aerial Monitoring Systems in Agriculture, Forestry, Defense, and Border Protection Using Artificial Intelligence (AI). In Alex Kang (Ed.), *Agriculture and Aquaculture Applications of Biosensors and Bioelectronics* (pp. 389–404). IGI Global.
15. Hernández, D., Cano, J. C., Silla, F., Calafate, C. T., & Cecilia, J. M. (2021). AI-enabled autonomous drones for fast climate change crisis assessment. *IEEE Internet of Things Journal*, *9*(10), 7286–7297.
16. Bilal, M., Rubab, F., Hussain, M., & Shah, S. A. R. (2023, November). Agriculture Revolutionized by Artificial Intelligence: Harvesting the Future. In *Biology and Life Sciences Forum* (Vol. 30, No. 1, p. 11). MDPI.
17. Ameer, S. A., Alkhafaji, M. A., Jaffer, Z., & Al-Farouni, M. (2024). We empower Farmers with IoT, UAVs, and Deep Learning in Smart Agriculture. In *E3S Web of Conferences* (Vol. 491, p. 04007). EDP Sciences.
18. Prakash, L. N. C. K., Ravva, S. K., Rathnamma, M. V., & Suryanarayana, G. (2023). AI applications of drones. In Sachi Nandan Mohanty et al. (Eds.), *Drone Technology: Future Trends and Practical Applications*, Chapter 7, pp. 153–182.

19. Subeesh, A., & Mehta, C. R. (2021). Automation and digitization of agriculture using artificial intelligence and the Internet of Things. *Artificial Intelligence in Agriculture*, *5*, 278–291.
20. Kaliannan, V., & Al Saidi, F.K.S. (2024). Intelligent Computing with Drones and Robotics for Precision Agriculture. In: Balasubramanian, S., Natarajan, G., Chelliah, P.R. (eds) *Intelligent Robots and Drones for Precision Agriculture. Signals and Communication Technology*. Springer, Cham.

12 Unveiling the Power of Digital Twins in Cognitive Farming

Kamaraj Arunachalam and
Senthil Kumar Jagatheesaperumal

12.1 INTRODUCTION

With the release of David Gelernter's *Mirror Worlds* in 1991, the concept of digital twin technology was first introduced. But it was Dr. Michael Grieves, who was teaching at the University of Michigan at the time, who is recognized for having introduced the idea of digital twin software and used the concept of digital twins in manufacturing for the first time in 2002. Finally, in 2010, NASA's John Vickers coined the phrase "digital twin." Indeed, it is accurate to say that NASA invented digital twin technology in the 1960s during its space exploration missions, when each traveling spacecraft was precisely duplicated in an earthbound version that NASA employees on flight crews used for research and simulation [1].

12.1.1 Defining Digital Twins in Agriculture

Digital twins come in several varieties according on how much a product is magnified. The field of application is where these twins diverge the most. It is typical for various kinds of digital twins to coexist in a system or procedure to be followed as [1]: 1) Component twins or Parts twins; 2) Asset twins; 3) System or Unit twins; 4) Process twins. Historically, agriculture has crossed four eras, and now it is in the fifth era. The concept and implementation of the technology in agriculture has tremendous growth. The generations of agriculture are as follows [2]:

- Agriculture 1.0 (till 1920): labor-intensive manual labor and animal traction;
- Agriculture 2.0 (1920–1960): motorization, the introduction and spread of tractor advancements (three-point hitch, diesel engine, tyres), but the importance of the manual workforce remains;
- Agriculture 3.0 (1960–1980): mechanization, genetic advancements, tractors with increased power and capacity to operate machinery for all manufacturing tasks, and the quick replacement of the manual labor force;

DOI: 10.1201/9781003487005-12

- Agriculture 4.0 (precision agriculture) (1980–2000): enhancement of machinery operation, installation of electronic control systems on tractors, and initial attempts at computerized/digitized farm management;
- Agriculture 5.0 (smart agriculture) (after 2000): integration of electronics and automation, specifically in mobile user-point procedures; spread of sensors for activity tracking and on-board GPS in tractors.

12.1.2 The Role of Digital Twins in Agriculture

Farmers may anticipate probable deviations, simulate the effects of interventions, and effectively respond to prospective concerns (or prevent them completely) by analyzing consolidated data from a single digital twin platform. It is not necessary to conduct an immediate on-site examination in order to do any of these actions remotely [3].

A constant monitoring and reporting system that notifies you of plant damage caused by pests, a cow's illness, equipment failure, dry soil, or a reduction in temperature is provided by digital twins. Additionally, digital twins are an effective tool for modelling and forecasting future events that can help you anticipate how farming practices and natural disasters might affect your company and plan ahead. The schematic view of digital twins is shown in Figure 12.1.

If digital twin technology is applied properly, it can give farmers unmatched insights and foresight into their land. Here's how to embrace the revolutionary effects of digital twins in farming step-by-step: 1) Real-time data collection; 2) Building a digital duplicate; 3) Data integration; 4) Observation and analysis; 5) Prediction and simulation; and 6) Optimization and decision-making

The following are the essential characteristics of Digital Twins [4],

- *Timeliness*: Since a digital twin is a reflection of its physical twin, changes in the physical object's state are detected and synchronized with the digital twin almost instantaneously.
- *Fidelity*: A digital twin's security and dependability must be beyond reproach, enabling decision-makers to put their whole faith in them.
- *Integration*: A digital twin ensures convergence in a consistent format by integrating data from several physical object features.

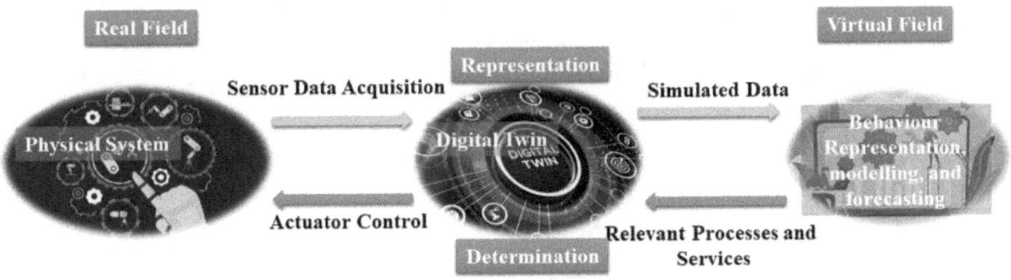

FIGURE 12.1 Symbolic illustration of digital twin.

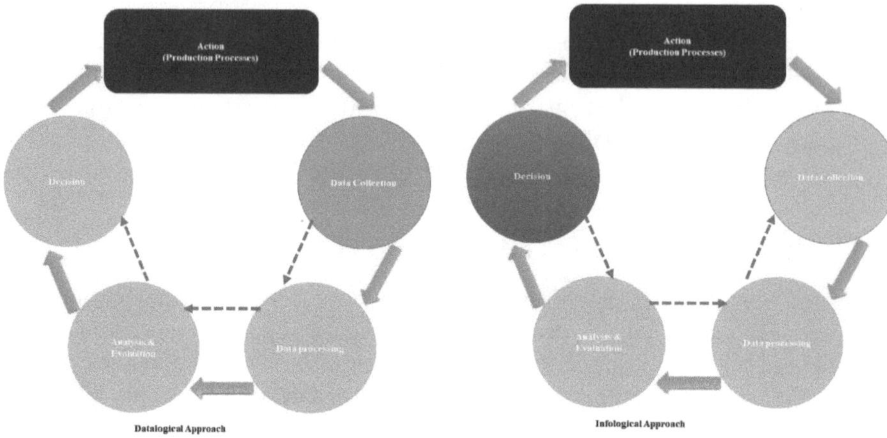

FIGURE 12.2 Two kinds of approaches: i) Datalogical approach and ii) Infological approach.

Also, it is expected to satisfy the following criteria, according to Dr. Sirko Schindler . A digital twin's needs for agriculture have to satisfy the following [5],

Data informed	It must be updated via digital measurement of the real-world thing.
Realistic	It must be a realistic surrogate for real-world thing.
Actionable	Information from real-world thing must be converted in to action

Approaches:

Figure 12.2 shows the datalogical (data-oriented) vs. infological (decision-oriented) approaches when designing information systems to support smart agriculture. The grey arrows indicate the flow with which (raw) data is transformed into information and decisions. The colored arrows, on the other hand, indicate the design logic of the information system. In the data-oriented approach (orange arrows) designers start by deciding what data to collect and then determine what decisions can be made [1].

In the alternative decision-oriented approach (green arrows), designers start by considering what are the problems to solve (i.e. what decisions to make) to finally determine what data to collect in order to carry out the entire decision-making process. The datalogical approach, although less rational, is often more natural to adopt since the flow of the project's logic proceeds in the same direction as the data-information transformation, as shown in Figure 12.3.

12.1.3 Overview of the Chapter Structure

Section 12.1 of the research chapter defined the digital twins, and agricultural eras, according the literature. Then it revealed the role of digital twins in agriculture and

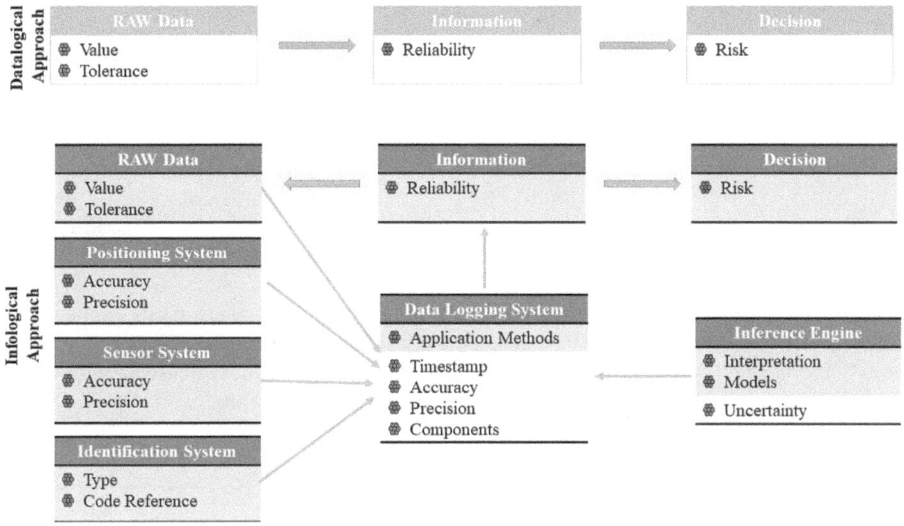

FIGURE 12.3 Detailed view of i) Datalogical approach and ii) Infological approach.

its characteristics. Thereafter it disclosed the two common approaches in information processing. Section 12.2 illustrates the foundations of cognitive farming, its concepts, key benefits, principles, and innovation domains, then it discusses integration of data analytics and artificial intelligence and the role of sensor technologies and IoT in agriculture. The advantages and limitations of IoT are also revealed. Section 12.3 enumerates the components and architecture of digital twins, types of applications in cognitive farming, with various types of twins. Section 12.4 explores two case studies in Kenya and Ethiopia. Section 12.5 provides n aenhanced view of challenges, opportunities, and future directions. Finally it is concluded with SWOT analysis.

12.2 FOUNDATIONS OF COGNITIVE FARMING

The intention is to enable decision support by enabling the examination of extremely intricate relationships between the biosphere and production to be used in a digital ecosystem of networked data and services, or "Agricultural Data Space" [6]. Cognitive farming includes types of sensors, wireless connectivity, livestock stock controls, and historically collecting and storing information [7]. Table 12.1 provides a summary of essential components and their significance.

Data modeling, data application, and data collecting are the three main parts of a digital twin. In order to achieve these goals, digital twins often make use of four essential technologies that can be combined to collect and store data in real-time, extract insightful information, and create a digital twin of the physical entity. They are 1) Internet of Things (IoT); 2) Cloud Computing; 3) Artificial Intelligence (AI); and 4) Extended Reality [9, 10].

TABLE 12.1
Summary of Essential components and its Significance

Concept	Key Components and Benefits
Soil–water	Irrigation planning, precision irrigation, and a decrease in crop yield losses
Irrigation	Hydroponics integrated with urban areas and forecasting models
Vertical farming	Monitoring and optimising the agri-food lifecycle, identifying forecasting
Robot	Robot selection, navigation, and analysis and performance evaluation
Crop	Calculating the expected yield and growth period of plants
Crop/hydroponic farm	Identification of factors that affect crop growth, such as ventilation systems, illumination, and outside temperature

Source: [8].

12.2.1 Principles and Concepts of Cognitive Agriculture

One of the main subfields of AI is cognitive computing, and scholars in this subject draw inspiration for their work from human cognition [11]. Also, cognitive computing is the simulation of human thought processes using computer models. The concept of human cognition, which is defined as the capacity of people to employ their five senses—vision, sound, smell, tongue, and touch—and react appropriately, is where the word cognition primarily originates.

Additionally, cognition is the capacity of people for independence, problem-solving for autonomous, adaptive, and anticipatory behavior [12]. These days, the features of integrating cognition into systems are changing due to advances in artificial intelligence. In the fields of computer science and artificial intelligence, there are sophisticated instruments, ideas, and technologies, but there is still a need to fully integrate cognition into these artifacts [13].

Decisions about these crucial elements will be supported by data from the agricultural enterprise that is digitally accessible. To achieve this, in order to make sustainable, fact-based decisions about crop production and livestock farming, data on factors related to the agricultural enterprise and the environment must be automatically collected using cutting-edge sensor technology and must be evaluated, analyzed, and processed cognitively in agricultural work and business processes as shown in Figure 12.4.

Domains of Innovation [14]

i. Data Space for Agriculture

Simple data interchange within a digital ecosystem tailored to agriculture that supports cognitive services and allows for the multivalent usage and connecting of large amounts of complicated agricultural data in safe data spaces.

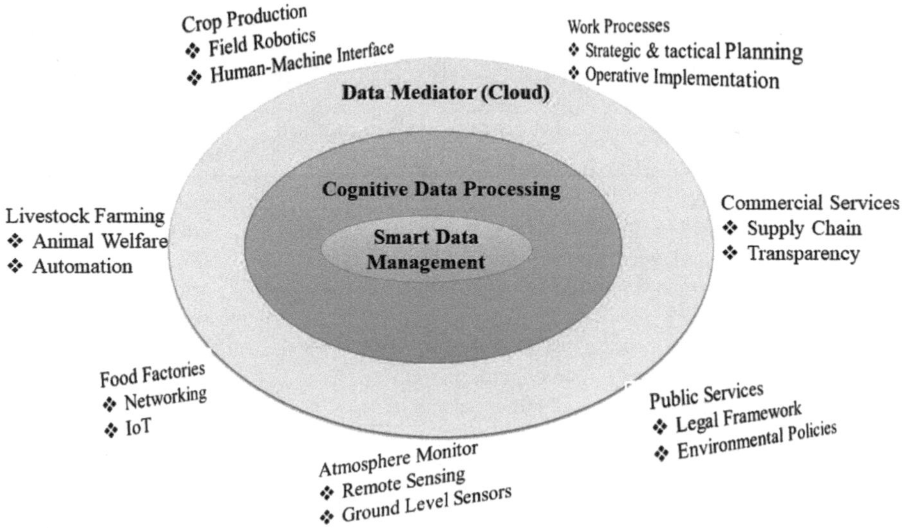

FIGURE 12.4 Core Digitalization Solutions.

ii. Innovative Sensor Technology

Automated multi-channel measurement data collection, integration, and interpretation from innovative robot and drone-assisted sensor systems to facilitate information-based decision-making.

iii. Creative Automation Ideas

A lightweight, electrified robotic platform with specialized sensor systems, autonomous field robotics is used for plant-specific field activities.

12.2.2 Integration of Data Analytics and Artificial Intelligence

Various kinds of sensors can be used in the agriculture field. Table 12.2 provides the information regarding methods available for irrigation automation [15].

Automation can be applied to irrigation, weeding, crop spraying, crop monitoring, and so on. In order to monitor and perform necessary actions the appropriate modeling is essential. For example, for weed monitoring, the modeling is being carried out [15] as shown in equation (1);

a) Destroyed weed rate = $(N_K/N_W) \times 100$
b) Correct spray rate = $(N_{CSR}/N_{SNWB}) \times 100$
c) False spray rate = $(N_{FSB}/N_{SB}) \times 100$
d) Herbicide reduction rate = $(1-N_{SB}/N_B) \times 100$ --- (1)

TABLE 12.2
Irrigation Automation Using Various Artificial Intelligence Technologies

S. No.	Algorithm / Method	Evapotranspiration method and desired calculation	Other techniques
1.	PLSR and other regression Algorithms	Evapotranspiration method	Data collecting sensors and IOT hardware
2.	Artificial Neural Network based control system	Evapotranspiration method	Sensors to measure temperature, wind speed, soil, etc.
3.	Fuzzy Logic	FAO Penman-Monteith method	Zigbee, WSN
4.	ANN (multilayer neural model),	Penman–Monteith method	-
5.	Fuzzy Logic Controller	Penman–Monteith method	Wireless sensors

Where, N_K = number of weeds killed; N_W = total number of weeds in the block; N_{CSR} = number of sprayed weed blocks; N_{FSB} = the number of sprayed non-weed blocks; N_{SB} = total number of sprayed blocks; N_B = total number of blocks examined. Similarly for the sprayer the following modeling has been performed as shown in equation (2),

a) Spray proficiency (%) = Minimum spray volume required × 100%
b) Application Rate in litre per Acre or Hectare = (Constant value of 495 or 600 British Matric × Nozzle Discharge Rate in litre every Minute) /(Effective Swath Width in Feet or Meters × Spraying speed in Miles or Kilometers every hour) --- (2)

Limitations [16]:

1. The major obstacle to the widespread use of AI in agriculture is the lack of straightforward solutions that smoothly integrate and embed AI in agriculture. Most farmers don't have the time or digital know-how to independently research AI solutions.
2. AI is limited to agriculture since it is unable to function outside of its programming.
3. A substantial amount of data is needed for AI systems to train their computers and generate reliable predictions.

Understanding the effects of environmental factors on yield, water resources, and resistance to recurrent exposure to biotic and abiotic stressors requires an understanding of both their spatial and temporal changes. For agricultural management (sowing time, crop rotation, crop choice, irrigation, etc.) and policy-making, a GIS-based tool called a decision support system will be useful. Researchers used a variety of data analysis methods to examine multi-scale environmental data for agricultural purposes [17].

Principal component analysis, auto correlation techniques, spectral analysis, parametric and non-parametric methods, and clustering method. Every one of these approaches has a unique benefit when it comes to comprehending how regional environmental parameter variability affects agricultural productivity [18].

12.2.3 Role of Sensor Technologies and IoT in Agriculture

Numerous sensors are needed to have well sophisticated digital twins in the agricultural field. The significant and most essential sensors and their functioning in agriculture are listed in Table 12.3 [19].

Agricultural sensors for intelligent farming

The introduction of sensors and the IoT to the agriculture industry could drastically alter the way we raise our food. These technologies enable farmers to collect and analyze data on a variety of crop-impacting environmental factors in real time, such as temperature, wind speed, soil moisture, smoke, flame, etc. Using smart agriculture sensors, farmers can keep an eye on the health of their crops, identify areas that need extra care, and schedule the planting, irrigation, fertilization, and harvesting of their crops [20].

Moreover, IoT and intelligent farming sensors can help farmers satisfy the growing need for food as the world's population continues to rise. Using these technologies can help farmers increase output while lowering costs and mitigating environmental effects. Conversely, wave-based technologies are finding quick use in smart sensors and industrial applications because of how well they analyze and handle complicated data. Wavelet methods are being more widely employed in many industries as helpful tools for industrial processes and modern sensor technologies because of their adaptability and efficiency in processing complicated data [21, 22]. Numerous mobile

TABLE 12.3
Essential Sensors and its Functioning in Agriculture

Sensor Name	Sensor Application
Soil NPK Sensor	Fertilizer Administration.
Soil Moisture Sensor	Irrigation Monitoring Systems, Forecasting and Harvesting.
Soil pH Sensor	Fertilizer Administration, Crop Disease Detection.
Electrical Conductivity Sensor	Fertilizer Administration, Quality Monitoring.
CO2 Sensor	Quality Monitoring, Processing Monitoring, Climate Conditions Monitoring.
Temperature and Humidity Sensor	Irrigation Monitoring Systems, Quality Monitoring, Processing Monitoring, Forecasting and Harvesting
Weather Station	Logistics Monitoring, Forecasting and Harvesting.
Water Level Sensor	Irrigation Monitoring Systems, Logistics Monitoring.
Livestock Sensor	Logistics Monitoring

TABLE 12.4
List of Mobile Apps for Agriculture Applications

S. No.	Category of Apps	Mobile Applications
1.	Agriculture management information Apps	Farm Manager, Agriapp, Bayer TurafXpert,Irrigo-crops, agriculture irrigation and cost Fertilizer removal by crop, Viewpoint, FieldView,OneSoil—Scouting—Satellite Field Monitoring
2.	Agriculture calculator Apps	Blend Calculator, Fertilizer Removal by crop Farm Calculators, Simple NPK. Calculator Corn N Rate Calculator, N Price Calculator, Fertilizer Cost Calculator, Illinois Manure Calculator
3.	News and Weather information Apps	Crop Monitor, AgWeb News & Markets, Farm Futures, Agri Zone: All in oneagri app, RealAg Markets, Ag Mobile
4.	Tank Mix Apps	Tank Mixing Ag app, TankMix, OnMRK, Agronote Farming Expenses
5.	Other Apps	Record Keeping Apps, Soil Sampling Apps, Precision Agriculture Apps, Area Apps, Commodity Pricing Apps

Source: [21].

applications are available to monitor and administer the agricultural needs. They are as follows in Table 12.4.

12.3 DIGITAL TWIN TECHNOLOGY IN AGRICULTURE

Digital twin technology is revolutionizing agriculture by creating virtual replicas of farming elements, enabling enhanced monitoring, analysis, and optimization. This section focuses on the fundamental aspects of digital twins, their architecture, and their diverse applications in cognitive farming.

12.3.1 Understanding Digital Twins

Digital twins are virtual versions of real-world objects with the ability to mimic, evaluate, and improve real-world operations. Digital twins are used in agriculture to mimic several aspects of farming, including soil, livestock, crops, and machinery [23]. This results in the creation of a coherent and dynamic digital ecosystem for modernizing agriculture. Digital twins give farmers a thorough and useful idea of their operations by utilizing real-time data from sensors and IoT devices [24]. Digital twins provide a revolutionary way to manage and optimize farming operations in the agricultural setting. They make it possible to precisely monitor the environmental factors, crop growth phases, and soil health, which enable farmers to customize their interventions for optimal production and efficiency [25]. Analogously, with real-time livestock monitoring, digital twins can identify early indicators of disease, thereby enhancing animal welfare and mitigating veterinary expenses [26]. All things considered, the application of digital twin technology to agriculture promotes a more productive and sustainable farming system by tackling the problems of contemporary agriculture with creative, data-driven solutions.

12.3.2 Components and Architecture of Digital Twin Systems

The main sources of data are sensors and IoT devices, which record data in real-time on a range of characteristics like crop health, soil moisture, nutrient levels, weather, and livestock behavior [27]. Modern sensors are able to measure a large number of different parameters, giving precise data that is necessary for modeling and simulation. Data storage and integration in part entails combining data from multiple sources into a single, central repository. To manage massive amounts of data and make sure it is available for analysis, cloud-based storage systems are frequently utilized [28].

Analytical models process and analyze the gathered data using methods from artificial intelligence and machine learning. They support decision-making by seeing trends, forecasting results, and offering insightful information [29]. An essential element that gives farmers a visual and interactive platform to monitor and manage their digital twins is the user interface. Dashboards let consumers make fast, well-informed decisions by displaying real-time data, analytics, and alarms [23]. Because of the user-friendly interface's design, even people with little technical knowledge may efficiently browse and operate the system. Sophisticated technologies like 5G and Low Power Wide Area Network (LPWAN) are important for reliable and effective data transfer [23].

A digital twin system is composed of multiple layers [30], where the physical assets, including machinery, crops, soil, animals, and cattle, as well as the connected sensors and IoT devices that gather data from these assets, are included in the physical layer.

12.3.3 Types and Applications in Cognitive Farming

Digital twins are essential tools for increasing the productivity and efficiency of farming operations. They can be grouped according to the various agricultural areas in which they are applied. Several noteworthy varieties and their uses consist of:

12.3.3.1 Crop Monitoring Twins

Crop monitoring digital twins accurately monitor a range of data, including growth phases, health, and yield estimates, by simulating the growth and development of actual crops. They monitor soil moisture, nutrient levels, and ambient variables with the use of sensor data [31].

12.3.3.2 Livestock Management Twins

The goal of livestock management twins is to replicate individual animals or entire herds in order to track productivity, behavior, and overall health. Farmers may better manage feeding and breeding programs, track vital signs, and identify illnesses early by combining data from environmental sensors and wearable devices [26]. Productivity rises and animal welfare is enhanced as a result.

12.3.3.3 Soil Management Twins

Twins in soil management examine the composition, fertility, and moisture content of the soil. Farmers may make educated judgments about crop rotation, fertilizer, and irrigation by regularly assessing the health of their soil [32].

12.3.3.4 Farm Machinery Twins

Tractors, harvesters, and irrigation systems are examples of agricultural machinery, and farm machinery twins mimic its functionality and maintenance requirements. Farmers may minimize downtime, increase equipment lifespan, and optimize maintenance schedules by keeping an eye on usage trends and mechanical health [24].

12.3.3.5 Resource Optimization Twins

The twins of resource optimization examine the economical use of resources such as energy, water, and fertilizers. To make sure that resources are used as efficiently as possible and to lessen waste and environmental effect, they examine data from a variety of sources [24].

12.4 CASE STUDIES: DIGITAL TWINS IN AGRICULTURE

This section provides detailed case studies of successful digital twin implementations in agriculture, highlighting diverse applications, challenges faced, and lessons learned.

12.4.1 Case Study 1: Kenya Agricultural Observatory Platform (KAOP)

The Kenya Agricultural & Livestock Research Organization (KALRO) framed the Kenya Agricultural Observatory Platform (KAOP) as a cutting-edge digital project. KAOP uses real-time data from sensors and IoT devices around the nation to generate virtual versions of the actual farming environment through the use of digital twin technology [33]. Through the creation of agronomic advisories for specific wards in the counties of Machakos and Kakamega, KAOP provides advices that are customized to the unique conditions of each stage of farming [34].

Among KAOP's accomplishments is timely planting advice, effective mitigation of climate change issues, and historical data that aids in farmers' comprehension of weather trends throughout time [35]. Farmers have been able to increase agricultural yields, optimize planting schedules, and lessen the negative effects of weather variability thanks to these advisories. Enhancing the precision of seasonal and short-term forecasts, as well as mapping every county in Kenya for more accurate data based on ecological zones and value chains, are concerns related to sustainability.

12.4.2 Case Study 2: Digital Agricultural Advisory Services (DAAS) in Ethiopia

The DAAS project tackles issues with Ethiopia's dairy industry, as native breeds produce inadequate amounts of milk. Through the use of digital twin technology, the project builds virtual dairy farm models that incorporate information on environmental factors, animal health, and feeding procedures [36]. The project, which is spearheaded by PxD Ethiopia and involves government agencies, academic institutions, research centers, and development partners, replicates individual cows and herds using digital twins [37].

The implementation in [38] is tailored to the requirements of certain virtual pig houses, their reproductive cycles, and local conditions, guaranteeing its applicability and providing useful insights. The project recognizes the important role that women farmers play in dairy production and promotes joint listening to advisory calls in order to alleviate gender-related inequities. By ensuring that both men and women may benefit from the advisory services, this inclusive strategy serves to improve overall dairy sector performance.

12.5 CHALLENGES, OPPORTUNITIES, AND FUTURE DIRECTIONS

This section explores the obstacles faced in implementation, potential areas for innovation, and future trends that could revolutionize farming practices.

12.5.1 Challenges in Implementing Digital Twins

Several obstacles stand in the way of using digital twin technology in agriculture. The high initial cost of deploying sensors, IoT devices, and related infrastructure is one of the main problems. Without outside financial assistance, small-scale farmers in particular could find it challenging to invest in these technologies. Furthermore, integrating many data sources and guaranteeing data veracity can be challenging, necessitating advanced data management tools and knowledge.

The requirement for strong communication networks is another major issue, particularly in rural locations where connectivity may be spotty or inconsistent. Digital twin effectiveness depends on real-time data transmission and synchronization between digital and physical entities. In addition, issues with data security and privacy exist. Farmers must have peace of mind knowing that their data is shielded from breaches and unwanted access.

12.5.2 Opportunities for Innovation and Improvement

Digital twin technology offers a lot of chances for innovation and advancement in agriculture, despite certain difficulties. The creation of affordable sensors and

IoT devices presents a substantial opportunity to increase the accessibility of technology for small and medium-sized farms. Digital twins' analytical powers can be further improved by developments in AI and machine learning, yielding more precise and useful findings. Innovation must be fueled by cooperation between farmers, legislators, agricultural specialists, and technology developers. Partnerships between the public and private sectors can supply the funding and resources required for research and development, as well as for pilot projects that highlight the benefits of digital twins in agriculture.

Furthermore, by minimizing environmental effect and optimizing resource consumption, the adoption of digital twins can support sustainable farming techniques. For example, focused fertilization can reduce chemical use and increase soil health, while precision irrigation control can conserve water. These enhancements may help agricultural systems remain resilient and sustainable over the long run.

12.5.3 Future Trends and Potential Impacts

The use of digital twin technology in agriculture is anticipated to be shaped by a number of trends in the near future. Digital twins will probably be used more widely as sophisticated sensors and IoT devices become more accessible and affordable. The stability and real-time capabilities of digital twins will be greatly improved as connectivity increases, especially with the rollout of 5G networks.

AI and machine learning will develop further, offering ever-more-advanced instruments for decision-making and data analysis. Thanks to these developments, digital twins will be able to provide prescriptive and predictive analytics, which will assist farmers in anticipating problems and proactively optimizing their operations. The summary of SWOT analysis of Digital Twins in agriculture field is provided in Figure 12.5.

Food security, resource scarcity, and climate change are some of the most urgent issues facing agriculture that digital twins may help with by improving production, efficiency, and sustainability. They may also provide farmers with the information and resources they need to make wise decisions, which will eventually result in farming communities that are more resilient and successful.

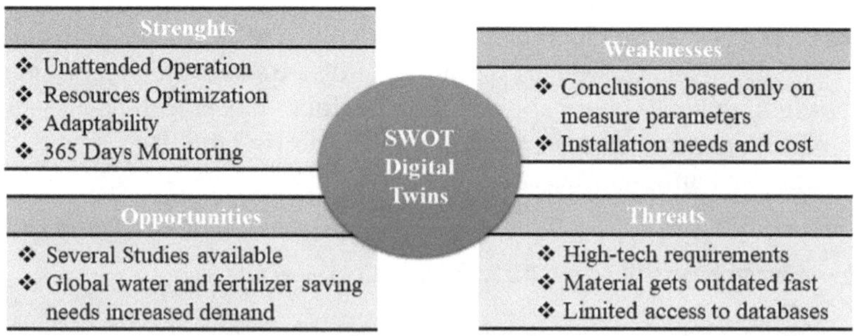

FIGURE 12.5 SWOT analysis of digital twins in agriculture.

12.6 CONCLUSION

Agriculture is about to undergo a transformation because of digital twin technology, which offers an advanced and integrated method of managing farming activities. Digital twins provide real-time monitoring, analysis, and optimization of physical assets by generating virtual replicas of them. This leads to increased productivity, sustainability, and efficiency. Cost, data integration, connection, and user training are some of the obstacles, but overall, the advantages might be substantial. There are lots of opportunities for innovation, especially with the developments in AI, ML, and IoT devices. Future-focused, the broad use of digital twins can help with pressing agricultural problems including food security, climate change, and resource scarcity. Through the promotion of stakeholder engagement and research and development funding, the agricultural industry may effectively leverage the revolutionary potential of digital twins, laying the groundwork for a more robust and profitable farming industry in the future.

REFERENCES

1. Cesco, Stefano, Paolo Sambo, Maurizio Borin, Bruno Basso, Guido Orzes, and Fabrizio Mazzetto. "Smart agriculture and digital twins: Applications and challenges in a vision of sustainability." *European Journal of Agronomy* **146** (2023): 126809.
2. Mazzetto, Fabrizio, Raimondo Gallo, Michael Riedl, and Pasqualina Sacco. "Proposal of an ontological approach to design and analyse farm information systems to support precision agriculture techniques." In *IOP Conference Series: Earth and Environmental Science*, **275**, no. 1, p. 012008. IOP Publishing, Bristol (England), 2019.
3. Fabbri, Carolina, Bruno Basso, Marco Napoli, Anna Dalla Marta, Simone Orlandini, and Rafael A. Martinez-Feria. "Developing a tactical nitrogen fertilizer management strategy for sustainable wheat production." *European Journal of Agronomy* **144** (2023): 126746.
4. Alves, Rafael Gomes, Gilberto Souza, Rodrigo Filev Maia, Anh Lan Ho Tran, Carlos Kamienski, Juha-Pekka Soininen, Plinio Thomaz Aquino, and Fabio Lima. "A digital twin for smart farming." In *2019 IEEE Global Humanitarian Technology Conference (GHTC)*, pp. 1–4. IEEE, 2019.
5. Peters, Diana, and Sirko Schindler. "FAIR for digital twins." *CEAS Space Journal* **16**, no. 3 (2024): 367–374.
6. Angin, Pelin, Mohammad Hossein Anisi, Furkan Göksel, Ceren Gürsoy, and Asaf Büyükgülcü. "Agrilora: A digital twin framework for smart agriculture." *J. Wirel. Mob. Networks Ubiquitous Comput. Dependable Appl.* **11**, no. 4 (2020): 77–96.
7. Pathan, Misbah, Nivedita Patel, Hiteshri Yagnik, and Manan Shah. "Artificial cognition for applications in smart agriculture: A comprehensive review." *Artificial Intelligence in Agriculture* **4** (2020): 81–95.
8. Trigo, Ana, Ana Marta-Costa, and Rui Fragoso. "Principles of sustainable agriculture: Defining standardized reference points." *Sustainability* **13**, no. 8 (2021): 4086.
9. Nasirahmadi, Abozar, and Oliver Hensel. "Toward the next generation of digitalization in agriculture based on digital twin paradigm." *Sensors* **22**, no. 2 (2022): 498.
10. Purcell, Warren, and Thomas Neubauer. "Digital twins in agriculture: A state-of-the-art review." *Smart Agricultural Technology* **3** (2023): 100094.

11. Gamage, T. A., E. M. N. M. Ekanayake, R. M. N. M. Rajapaksha, M. T. D. Mackonal, and D. D. M. Ranasinghe. "Integrating artificial cognitive systems in smart agriculture." *15th International Research Conference of General Sir John Kotelawala Defence UniversityAt: Ratmalana, Sri Lanka* (2022): 89–94.
12. Tagarakis, Aristotelis C., Lefteris Benos, George Kyriakarakos, Simon Pearson, Claus Grøn Sørensen, and Dionysis Bochtis. "Digital twins in agriculture and forestry: A review." *Sensors* **24**, no. 10 (2024): 3117.
13. Alves, Rafael Gomes, Rodrigo Filev Maia, and Fábio Lima. "Development of a Digital Twin for smart farming: Irrigation management system for water saving." *Journal of Cleaner Production* **388** (2023): 135920.
14. Devitt, S. Kate. "Cognitive factors that affect the adoption of autonomous agriculture." *Farm Policy Journal* **15**, no. 2 (2021): 49–60. *arXiv preprint arXiv:*2111.14092
15. Talaviya, Tanha, Dhara Shah, Nivedita Patel, Hiteshri Yagnik, and Manan Shah. "Implementation of artificial intelligence in agriculture for optimisation of irrigation and application of pesticides and herbicides." *Artificial Intelligence in Agriculture* **4** (2020): 58–73.
16. Javaid, Mohd, Abid Haleem, Ibrahim Haleem Khan, and Rajiv Suman. "Understanding the potential applications of Artificial Intelligence in Agriculture Sector." *Advanced Agrochem* **2**, no. 1 (2023): 15–30.
17. Akter, Most Sohana. "Harnessing technology for environmental sustainability: utilizing AI to tackle global ecological challenge." *Journal of Artificial Intelligence General Science (JAIGS) ISSN: 3006-4023* **2**, no. 1 (2024): 61–70.
18. Linaza, Maria Teresa, Jorge Posada, Jürgen Bund, Peter Eisert, Marco Quartulli, Jürgen Döllner, Alain Pagani et al. "Data-driven artificial intelligence applications for sustainable precision agriculture." *Agronomy* **11**, no. 6 (2021): 1227.
19. Morchid, Abdennabi, Rachid El Alami, Aeshah A. Raezah, and Yassine Sabbar. "Applications of Internet of Things (IoT) and sensors technology to increase food security and agricultural sustainability: Benefits and challenges." *Ain Shams Engineering Journal* **15**, no. 3 (2023): 102509.
20. Rajak, Prem, Abhratanu Ganguly, Satadal Adhikary, and Suchandra Bhattacharya. "Internet of Things and smart sensors in agriculture: Scopes and challenges." *Journal of Agriculture and Food Research* **14** (2023): 100776.
21. Farooq, Muhammad Shoaib, Shamyla Riaz, Adnan Abid, Kamran Abid, and Muhammad Azhar Naeem. "A survey on the role of IoT in agriculture for the implementation of smart farming." *IEEE Access* **7** (2019): 156237–156271.
22. Farooq, Muhammad Shoaib, Shamyla Riaz, Adnan Abid, Tariq Umer, and Yousaf Bin Zikria. "Role of IoT technology in agriculture: A systematic literature review." *Electronics* **9**, no. 2 (2020): 319.
23. Peladarinos, Nikolaos, Dimitrios Piromalis, Vasileios Cheimaras, Efthymios Tserepas, Radu Adrian Munteanu, and Panagiotis Papageorgas. "Enhancing smart agriculture by implementing digital twins: A comprehensive review." *Sensors* **23**, no. 16 (2023): 7128.
24. Verdouw, Cor, Bedir Tekinerdogan, Adrie Beulens, and Sjaak Wolfert. "Digital twins in smart farming." *Agricultural Systems* **189** (2021): 103046.
25. Chaux, Jesus David, David Sanchez-Londono, and Giacomo Barbieri. "A digital twin architecture to optimize productivity within controlled environment agriculture." *Applied Sciences* **11**, no. 19 (2021): 8875.
26. Neethirajan, Suresh, and Bas Kemp. "Digital twins in livestock farming." *Animals* 11, no. 4 (2021): 1008.

27. Paul, Kenny, Sandeep S. Chatterjee, Puja Pai, Alok Varshney, Siddhi Juikar, Venkatesh Prasad, Bhaskar Bhadra, and Santanu Dasgupta. "Viable smart sensors and their application in data driven agriculture." *Computers and Electronics in Agriculture* **198** (2022): 107096.
28. Stergiou, Christos L., and Kostas E. Psannis. "Digital twin intelligent system for industrial IoT-based big data management and analysis in cloud." *Virtual Reality & Intelligent Hardware* **4**, no. 4 (2022): 279–291.
29. Kaur, Maninder Jeet, Ved P. Mishra, and Piyush Maheshwari. "The convergence of digital twin, IoT, and machine learning: Transforming data into action." *Digital Twin Technologies and Smart Cities* **2**, no. 1(2020): 3–17.
30. Redelinghuys, A. J. H., Anton Herman Basson, and Karel Kruger. "A six-layer architecture for the digital twin: a manufacturing case study implementation." *Journal of Intelligent Manufacturing* **31**, no. 6 (2020): 1383–1402.
31. Madeira, Rui Neves, Pedro Albuquerque Santos, Oskars Java, Torsten Priebe, Eduardo Graça, Eszter Sárközi, Bernward Asprion, and Raquel Pinto-Bello Gómez. "Towards digital twins for multi-sensor land and plant monitoring." *Procedia Computer Science* **210** (2022): 45–52.
32. Silva, Letícia, Francisco Rodríguez-Sedano, Paula Baptista, and João Paulo Coelho. "The digital twin paradigm applied to soil quality assessment: A systematic literature review." *Sensors* **23**, no. 2 (2023): 1007.
33. Kutyauripo, Innocent, Munyaradzi Rushambwa, and Lyndah Chiwazi. "Artificial intelligence applications in the agrifood sectors." *Journal of Agriculture and Food Research* **11** (2023): 100502.
34. Kasper, Lukas, Felix Birkelbach, Paul Schwarzmayr, Gernot Steindl, Daniel Ramsauer, and René Hofmann. "Toward a practical digital twin platform tailored to the requirements of industrial energy systems." *Applied Sciences* **12**, no. 14 (2022): 6981.
35. Gikunda, Kinyua. "Harnessing artificial intelligence for sustainable agricultural development in Africa: Opportunities, challenges, and impact." *arXiv preprint arXiv:2401.06171* (2024).
36. Dominic, Ditty Maria, Rasheed Sulaiman V., Nimisha Mittal, Sheetal Sharma, and Gakhar Shalini. "Study on Enhancing capacities of Extension and Advisory Services (EAS) in using data and information Available through digital platforms and applications in Odisha, India." *CGIAR Initiative on Digital Innovation* **1** (2023), 45.
37. Kropff, Wietske, Berta Ortiz-Crespo, Lidya Tesfaye, Sintayehu Alemayehu, Jonathan Steinke, Brenda Binge, Julian Ramirez Villegas et al. "Mainstreaming digital approaches for adaptation in agriculture in Ethiopia." *Global Center on Adaptation* (2023), 1–59.
38. Jeong, Deuk-young, Seng-Kyoun Jo, In-Bok Lee, Hakjong Shin, and Jun-Gyu Kim. "Digital twin application: Making a virtual pig house toward digital livestock farming." *IEEE Access* **11** (2023), 121592–121602.

13 Prospective Applications of Nanotechnology in Sustainable and Precision Agriculture

Sakthi P, Uma J, and Veeramani P

13.1 INTRODUCTION

The field of science that studies materials at the nanoscale and their applications is known as nanotechnology. It entails the synthesis, material characterization, and system design that reduces everything to the nanoscale of range 1–100nm. Nanotechnology has immense applications in various fields like engineering, biotechnology, medicine, defence, energy, environment and agriculture, etc. American physicist Richard Feynman introduced the concept of nanotechnology in a speech at the California Institute of Technology during the American Physical Society's annual conference in December 1959 [1]. The nanomaterials can be synthesized using top-down and bottom-up approach. The top-down method primarily relies on physical techniques to produce materials at the nanoscale, such as laser ablation, inert gas condensation, and ball milling. Bottom-up techniques involve chemical processes including chemical vapour deposition, sol-gel synthesis, and hydrothermal procedures. These methods use more hazardous and poisonous chemicals to create NPs. The disadvantage of this approach is that during the synthesis of nanomaterials, hazardous byproducts are released. Solution combustion synthesis is one perspective chemical technique that helps prevent these side effects. These days, a different synthesis technique called the biological method—which uses biological materials like plants, microbes, and algae to create NPs—is employed to get around these problems. This method's advantage is that it's inexpensive, comparatively toxic-free, and eco-friendly.

Agriculture has become essential in enhancing economic growth and in providing the nutritional needs of the people. A variety of elements, including weeds, insects and pests, population, and economic conditions, influence sustainable agriculture. Developing the agricultural aspect's grade is crucial to addressing the issues and needs. Large amounts of fertilizers and pesticides are used to increase the crop yield. The chemicals present in the fertilizers affect the soil which leads to its infertility. According to a number of statistics, the world uses two million tonnes of pesticides annually, with Europe using 45% of them and the USA using 25% [2]. As a result, the

DOI: 10.1201/9781003487005-13

first and most important action that has to be done is to manage resources properly and deal with any related problems. Consequently, academics and agriculturalists can cooperate with agri-nanotechnology as a useful instrument to raise the agri-food sector's standing internationally. Sustaining food security depends on the agriculture industry remaining stable. The only more effective options that researchers and agriculturists should use to preserve an economically and environmentally sustainable agriculture are nano pesticides, nano insecticides, fertilizers, fungicides, and herbicides: collectively, nano agrochemicals.

13.1.2 Technologies in Precision Agriculture

Information technology (IT) is utilized in precision agriculture (PA) to ensure that agricultural land and its yield obtain a precise amount of nutrition for its increased strength and production. Precision farming encompasses the utilization of hardware, such as GPS and GIS systems, as well as various GIS software and conventional agricultural management approaches. The precision agriculture cycle shown in Figure 13.1 has positive effects on agriculture, production, environmental and economic aspects. Considering the components and technology of precision agriculture enables farmers to make better decisions about their farming practices and specifications, enabling them to carry them out in the fields with the precise quantity of input needed. Precision farming minimizes input costs, increases crop yield, and lowers dangers to the environment and public health by using location-specific agricultural techniques. Agricultural rotation and irrigation techniques are used in order to increase agricultural yield while lowering input costs.

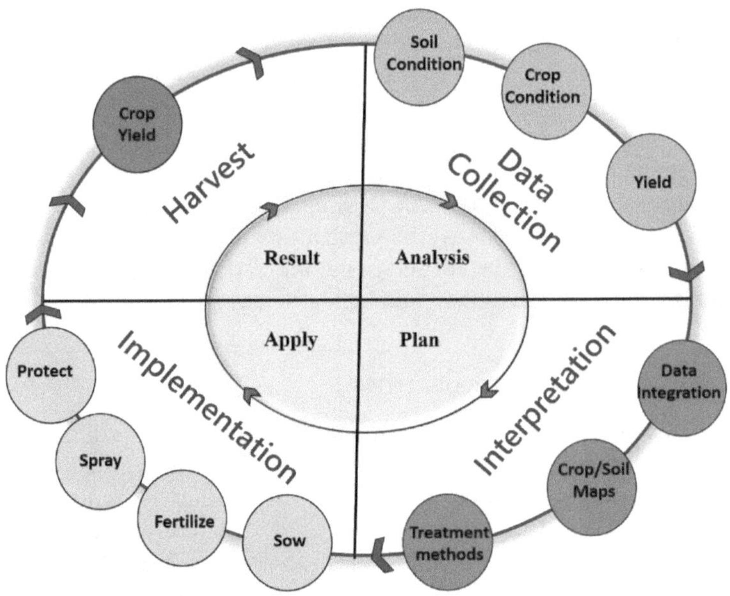

FIGURE 13.1 Precision Agricultural Cycle.

13.1.2.1 GIS and GPS

The entire agricultural system is reorganized with low input, high efficiency, and sustainable, precision agriculture methods are implemented. The development of multiple technologies such as the GPS, GIS, automation and control, automatic control, remote sensing, mobile computing, information processing, and telecommunications have been largely implemented to attain a precision agriculture. Predictive modelling and geographic analysis have taken the place of the farmer's almanac. Using these technologies, farmers can now manage their enterprises precisely by visualizing their cultivating land, yield condition, and management techniques in never-before-seen ways. Obtaining geographic data is becoming a necessary farming practice. Websites that provide useful information to guide the farmers in understanding their land better and taking more educated decisions are hosted by government.

A geographic information system (GIS) is a computer program that can be loaded on any modern desktop computer model and is intended to gather, store, organize, and present any kind of geographic data that is referenced by geographic coordinates. The major commercial drive towards precision farming is predicated on the integration of GIS and modern technology, which allows precise yield data that is geographically referenced, as well as variable rate applications of pesticides and fertilizers. Field data (slope, aspect, nutrients, and yield) can be recorded by the global positioning system (GPS) as geographical latitude and longitude data.

13.1.2.2 Remote Sensing

With this approach, precise data may be gathered at a distance without having to touch the thing. Because of the high spatial, temporal, and spectral resolution of remote sensing technologies and the widespread availability of satellites and unmanned aerial vehicles (UAV), these tools have gained popularity in agricultural situations. Through the conversion of visible and invisible field properties into spatial information, remote sensing dynamically monitors plant and soil conditions. The digital notepad and GIS database are georeferenced to the most widely used tool for visual observations. Data collection, transmission, processing, and storage systems are all part of the remote sensing system. The five primary components of remote sensing system are data transmission, data preprocessing, control and positioning system, carrier platform, and remote sensor. At the present time, UAVs and airships are used to capture the images. Some of the advantages of the UAVs are low flying costs, higher flying altitudes, increased flexibility, and high efficiency.

13.1.3 Nanoparticles in Agriculture

Nanotechnology is also being used in modern agriculture to realize the precision agriculture idea. Numerous uses for nanomaterials exist in agriculture, including nutrition, disease and weed control, and administration of farming techniques. In agriculture, nanoparticles fall into three categories: carbon-based, inorganic, and organic. Nanomaterials made by biological processes, such as plant extracts, have become an innovative way to reduce waste output and maintain the sustainability of ecosystems. Precision farming uses a variety of biological sources for nanoparticle

TABLE 13.1
Uses of Nanoparticles in Agriculture

Nanoparticles	Agricultural Area	Uses
Silver Nanoparticles	Plant Growth	Antimicrobial Activity
		Plant Growth
		Fertilizers
	Soil Improvement	Increase Soil fertility
		Water retention
	Plant Breeding	Genetic material for plant transformation
Zinc Oxide nanoparticles	Plant Growth	Improve seed germination
		Increase protein content
		Increase root growth
		Increase elongation of roots
		Antimicrobial agent
Titanium dioxide	Plant Growth	Promote Seed Germination
		Increase plant growth
		Increase crop yield
Carbon Nanotubes	Water Treatment	Water purification
	Nanosensors	Monitor plant growth and environment conditions

manufacturing. Some of the nanoparticles that are used in agriculture are discussed below in Table 13.1.

13.1.3.1 Titanium Dioxide

Outstanding with greater efficiency, photochemical stability, benign nature, and inexpensive price, titanium dioxide (TiO_2) has been recognized as the most effective and useful photo catalyst to date for both research and industrial applications. When TiO_2 is used in the proper concentration, it has a positive effect on the germination of seed. They accelerate nutrient intake, improve photosynthesis and chlorophyll content, boost the synthesis of certain enzymes and proteins, and boost stress tolerance [3]. The rate of photosynthesis was positively correlated with the observed increase in chlorophyll concentration, which led to enhanced plant development and increased yields in tomato plants [4]. The wide antibacterial activities of TiO_2 NPs demonstrate their exceptional inhibitory effect against a variety of bacterial and fungal pathogens.

13.1.3.2 Silver Nanoparticles

The most researched and used nanoparticle for biosystems is silver nanoparticles (AgNPs). They possess a wide range of antimicrobial activity and have been shown to have potent bactericidal and inhibitory effects. The tomato seeds treated with silver nanoparticles had a 70% higher germination rate and a shorter mean germination time. AgNPs are a powerful pest management tool with a variety of advantages, including non-toxicity, safety, and enhanced pest control techniques. Effectively giving the target plants a high and appropriate dose, silver nanoparticles act as the nano-based insecticide. The pesticidal activity of the AgNPs against Tribolium castaneum [5] was highest. The results of a bioassay involving ovipositional deterrence, antifeedent tests, mortality tests, and repellent activity indicate that malathion and silver nanoparticles exhibit the highest levels of pesticidal activity. Using polyurethane foam, the efficacy of silver nanoparticles as a pesticide against chlorpyrifos was studied. Between 92% and 94% of the pesticide was removed at the maximum rate of 20 mL/h [6].

13.1.3.3 Zinc Oxide Nanoparticles

Zinc oxide nanoparticles may increase crop growth and productivity. Zinc oxide nanoparticles were used in varying concentrations on the seeds of peanut. The application of 25nm mean particle size of ZnO nanoparticles at 1000 ppm concentration increased plant growth, seed germination, and seedling vigour. Zinc oxide nanoparticles are a novel form of zinc fertilizer that have been widely utilized in agriculture. Numerous research has been done to evaluate the effects of ZnO NPs on plant growth. ZnO enhanced the rice's quality by raising the head milling rate, decreasing the proportion of chalky grains, and raising the rice's breakdown and flavour value.

13.1.3.4 Carbon Nanotubes

Because of their potential to improve plant growth and germination of seeds, their capacity to pass through plant cell walls, and their ability to be delivered to individual cells, carbon nanotubes (CNTs) have generated a great deal of interest in the agricultural industry. Furthermore, it is mentioned that the CNTs can pierce the thick seed coat and initiate the process of water absorption. When the seeds of Brassica juncea(mustard) were soaked with 30nm MWCNTs, the time taken for 50% germination, showed increased shoot and root growth. It was noticed that the moisture content of the oxidized MWCNT-treated seeds was noticeably higher than that of the untreated condition, suggesting that the oxidized MWCNTs enhanced the seeds' capacity to absorb water and regenerate quickly. The facile penetration of the functionalized CNTs was the cause of the elevated water content in the seeds treated with oxidized MWCNTs [7]. Because of their low toxicity and ability to migrate through the cell walls of plants, cell membranes, and cell organelles, CNTs offer a great deal of promising nanomaterial as a carrier or transportation system for genomic material or nourishing material into the cell organelles like the vacuole and plastids [8].

13.1.4 Nanotechnology Applications in Agriculture

Crop development, production, processing, storage, packing, and transportation are just a few of the stages of agriculture where nanotechnology has a potential application. Agriculture may benefit from the easy production of special-featured nanoparticles from a variety of biological sources. Producing nanoparticles has proven to be a successful use of plant cuttings (e.g. leaves, floral parts, barks, and roots) from a broad range of plant species. Some of the prospective uses of nanotechnology in farming discussed below are shown in Figure 13.2.

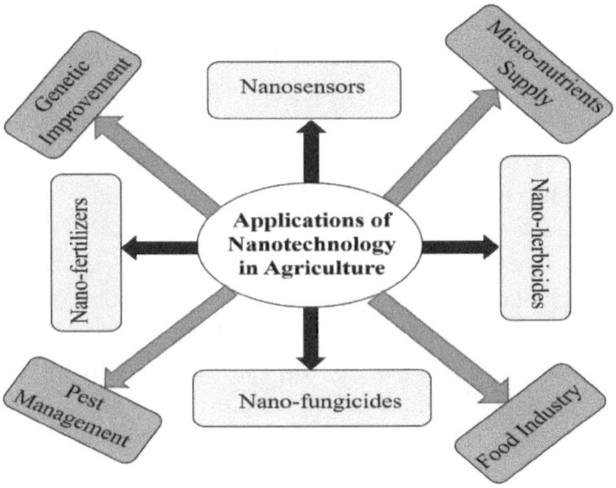

FIGURE 13.2 Prospective Applications of Nanotechnology in Agriculture.

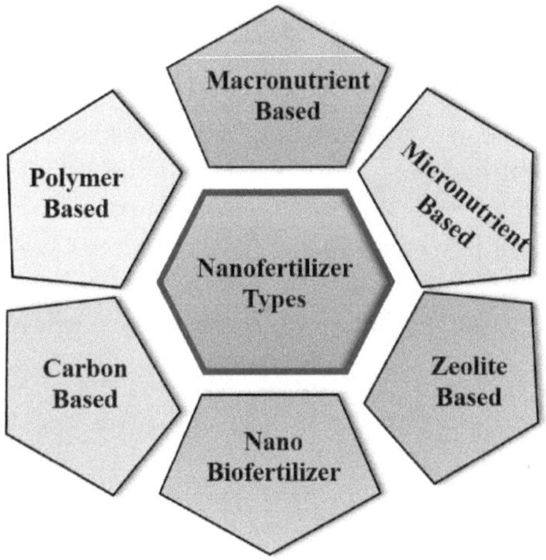

FIGURE 13.3 Nanofertilizer Types.

13.1.4.1 Nanofertilizers

Scientists are currently researching on creating chemical fertilizers with a delayed release rate. For instance, hydroxyapatite and urea can be combined to create a fertilizer with a slow-release rate that releases nutrients for plants gradually. Furthermore, the release of surplus nutrients has an adverse effect on the environment, which has prompted the creation of more environmentally friendly and efficient fertilizers. With their increased effectiveness and decreased environmental impact, nanofertilizers have emerged as a viable answer to these problems. Nanofertilizers can be categorized according to their consistency, activity, and nutrient makeup. The nanofertilizers mainly classified according to the type of material are shown in Figure 13.3.

13.1.4.2 Macronutrient Nanofertilizers

Macronutrient nanofertilizers have a number of advantages for plant growth and environment sustainability. For example, the nitrogen-based nanofertilizers increase nitrogen consumption efficiency and reduce the release of greenhouse gases and eutrophication [9]. Phosphorus nanofertilizers have shown promising results in soil reclamation, in addition to helping plants absorb nutrients. The physical features of the soil are enriched by the potassium-based nanofertilizers which have more absorption capability and are more resistant to leaching. The benefits of calcium-based nanofertilizers include improved fruit and vegetable quality, increased crop yields, and increased resistance to disease. Magnesium-based nanofertilizers benefit a range of kinds of crop by enhancing crop quality, growth, and pest and disease resistance. Soil acidification was reduced by the supply of nutrients with the sulfur-based nanofertilizers which slowly release sulfur into the soil. Nitrogen molecules are mixed with NPs such metal oxides, graphene, and carbon nanotubes to create nitrogen nanofertilizers. By increasing the amount of nitrogen that is available in the soil, this mixture of particles enables plants to absorb more nutrients. According to one study, which looked at the nutrient release patterns of nitrogen-containing nanofertilizer formulations, standard fertilizer only lasts for 300–350 hours while nanofertilizer delivers nutrients for up to 1200 hours [10].

13.1.4.3 Micronutrient Nanofertilizers

Advanced agricultural inputs known as micronutrient-based nanofertilizers use NPs to supply vital micronutrients to plants more efficiently than traditional fertilizers. Although they are needed in trace amounts, micronutrients including B, Cu, Fe, Ni, Zn and Ti are essential for plant growth and development as well as general health. As will be covered below, micronutrient nanofertilizers have several benefits for plants, such as better nutrient intake, higher yield, and increased resilience to biotic and abiotic stressors shown in Figure 13.4.

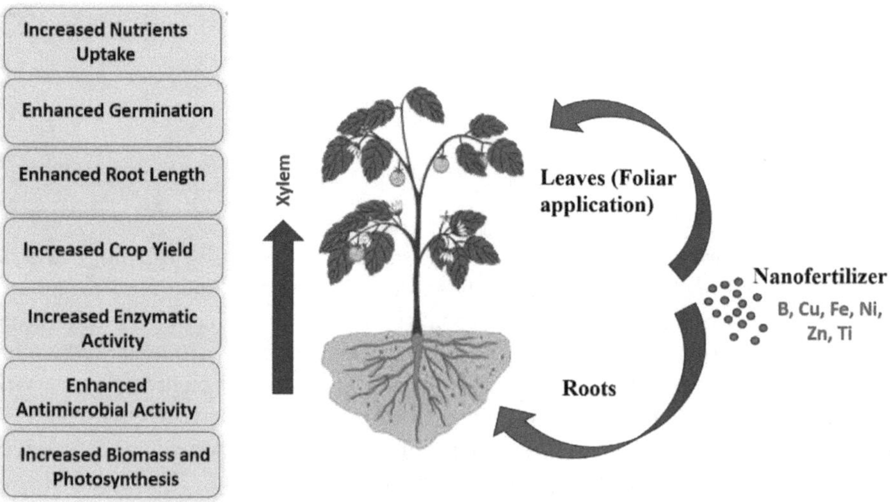

FIGURE 13.4 Micronutrient Nanofertilizer Application to the Plant.

Iron, especially heme proteins, is necessary for the electron transport system, chlorophyll production, and some enzyme activities. Zinc is needed by a variety of metabolic enzymes for a variety of catalytic functions and processes, including photosynthesis, tryptophan synthesis, protein synthesis, cell division, membrane potential, and structure maintenance. Plant cells can be penetrated by the boron-based nanofertilizers, which supply them with vital micronutrients. Copper is a naturally occurring antibacterial and antifungal agent, that aids plants in resisting pests and illnesses [11]. Numerous metabolic functions, such as photosynthesis, respiration, and nitrogen metabolism, depend on nickel. Also, titanium-based nanoparticles offer long-term protection against illness and pests because of their antibacterial qualities. These NPs were found to be beneficial in lowering soil salinity when applied to field-grown broad bean crops.

13.1.4.4 Polymer Nanofertilizers

Many studies are being conducted on the use of polymers as nanofertilizers. Chitosan is a cheap, cationic biopolymer that is used to increase agricultural output, photosynthetic rate, nutrient uptake, plant growth, and seed germination. Nanogels are porous structures made of a polymer blended with a liquid; they have the ability to absorb fertilizers or other substances and release them gradually over time. The essential ingredients for promoting plant growth—nitrogen, phosphorus, and potassium—are infused into the nanogels. A new class of fertilizers known as polyurethane-based nanofertilizers takes advantage of the properties of polyurethane nanoparticles (NPs), which are made of organic units joined by urethane linkages [12]. In addition to offering regulated nutrient release into the soil, the polyurethane matrix of these nanofertilizers protects the particles from the external environment.

13.1.4.5 Carbon based Nanofertilizers

Since carbon is a component of all organic substances and is involved in biological pathways, it is vital for supporting life. Additionally, plants can benefit greatly from nanostructured carbon [13], as demonstrated by the use of biochar, a byproduct of residual biomass from field plants. Many carbon nanostructures found in the biochar oxidize when exposed to air, forming pores on the layers. These pores assist in keeping the soil moist and nourished for germinating seeds and plant growth. A carbon nanotube (CNT) is a synthetic carbon nanomaterial with remarkable stability and physicochemical characteristics. Growing plants in CNT-enriched nutritional media lengthened their shoots and promoted seed germination.

13.1.4.6 Zeolite Based Nanofertilizers

Zeolite-based NFs improve the intake of nutrients by the crops during the plant development cycle with slow release of nutrients to the plants. This stops the soil from losing nutrients as a result of nitrogen fixation, leaching, denitrification, and volatilization. The process of creating zeolite nanoparticles (NPs) [14], which are then mixed with different substances to form fertilizer, is the basis for developing zeolite-based nanofertilizers. Because these fertilizers are tiny, they can more effectively reach the root zone of a crop by penetrating deeper into the soil. It is well known that zeolite-based compound fertilizers improve lettuce plants' nutrient uptake efficiency.

13.1.4.7 Nano-fungicides

Fungicidal formulations containing fungicide particles smaller than 10^{-9} are known as nano-fungicides. It is possible to create fungicidal nanoparticles with copper, silver, and other metals. Recent investigations have proven the microbicidal capabilities of AgNPs. The experiments show that deactivating ribosomal proteins, cell proteins, and the mitochondrial enzymes responsible for ATP synthesis can prevent DNA replication in pathogenic fungal microorganisms when treated with silver nanoparticles. When PVP (polyvinylpyrrolidone) was used to stabilize silver nanoparticles (AgNPs), the growth of Candida krusei, Candida albicans, Candida glabrata, Candida tropicalis, and Aspergillus niger was significantly inhibited [15]. AgNPs that employ the leaf extract of A. indica have demonstrated a remarkable ability to prevent the growth of certain fungi, including A. alternata, B. cinerni, and others [16].

Copper stands out as a popular solid antifungal agent among metallic nanoparticles. The scientists noted that the release of Cu^{2+} ions into the cellular atmosphere was the reason behind the antimicrobial properties of the nanoparticles. The Cu^{2+} ions that were released interact with the carboxyl and amino acid groups as well as the sulfhydryl groups of amino acids upon coming into touch with an outer bacterial cell membrane. This ultimately leads to conformational changes in the structure of proteins [17]. A 9-day incubation period with 450 ppm of Cu Nps suppressed 93.98% of fungal growth, according to the survey [18].

13.1.4.8 Nanoherbicides

Unwanted plants known as weeds damage the crop plants in agricultural fields. Certain chemicals, including herbicides, which are used to eradicate weeds, have an adverse effect on food crops. With no harmful residues left in the soil or surrounding ecosystem, nanoherbicides can be extremely helpful in completely eliminating weeds from crops in an environmentally responsible way. The chemical composition of the materials has led to the classification of nanoherbicides into the following three primary groups.

13.1.4.9 Organic Nanoherbicides

Organic nanomaterials can be based on polymers, synthetic organic materials, lipids, cellulose-based substances, proteins, and complex macromolecules like dendrimers. They are excellent building blocks for the creation of nanoherbicides. Polymers' being biodegradable, low cost, and biologically compatible make them popular choices for use in the manufacture of nanoherbicides [19]. Additionally, it has been reported that biopolymer-based nanocarriers for herbicide, like chitosan (CS), can be synthesized utilizing chemical crosslinking agents like tripolyphosphate (TPP) or functionalization with compounds like N-octyl derivatives or 11-mercapto undecanoic acid (MUA) [20]. Finding effective nano-enabled herbicides may also depend on chitosan functionalization and bioconjugation. When it comes to herbicide assembly, synthetic organic materials have a clear advantage over natural polymers due to their superior chemical and physical strength, hydroxide resistance, protection from erosion, and acid resistance.

13.1.4.10 Inorganic Nanoherbicides

For the purpose of increasing the crop yield and the performance of herbicides, which reduces waste and runoff, inorganic nano-enabled materials are being studied more and more for agricultural applications. Herbicides that are enabled by inorganic nanotechnology depend on silica, metal, mesoporous silica nanoparticles, etc. since inorganic nanomaterials may decrease herbicide leaching through soil and, therefore, increase the conveyance of the important functional ingredients to the plant. Inorganic materials have been widely used for herbicide alliance in the amalgamation of nanoherbicides based on double-layer aluminium hydroxides or zinc, or sepiolite clay linked with magnesium–aluminium. They fight Chlamydomonas reinhardtii Dang algae and help the system encapsulate hydrophobic pesticides among their layers [21]. Following the impacts on soil enzymatic activity, the stability of nano-encapsulated metribuzin in polymeric nanoparticles (nanoMTZ) was confirmed over time, along with its dissipation in various soils. On Ipomoea grandifolia plants, the physiological characteristics and control effects of nanoMTZ were examined [22].

13.1.4.11 Hybrid Nanoherbicides

The advantages of combining organic and inorganic materials can be integrated into a single assembly by using a hybrid nanomaterial herbicides synthesis process. These adaptable nanomaterials come in a variety of forms, sizes, morphologies, and chemical compositions. Furthermore, hybrid nanoherbicides can provide excellent targetability, traceability, and stimulus responsiveness characteristics. Furthermore, new structures known as metal-organic frameworks (MOFs) possess several attributes, including large surface areas, high stability, and variable hybrid compositions that make them appropriate carriers for the precise release of herbicides [23]. Since MOFs' transport mechanism can make it easier for plant cells to absorb the herbicides, the herbicides encapsulated in them have a stronger herbicidal activity than the active ingredient alone.

13.1.4.12 Nanosensors

Nanobiosensors are the next phase of the biosensor technology to sense the physical and chemical characteristics of the medium. The nanobiosensors' nanosized particles function as bioreceptors on transducers. The signal to identify one (or more) fragments is supplied to recognition elements. The intriguing features that include the biocomponents of a transduction system in an intricate design to enhance the systematic performance of nanomaterials are immobilization, miniaturization, and fictionalization [24]. Recently many nano-based biosensors have been developed to monitor the plant growth, soil quality and to detect the plant pathogens shown in Figure 13.5. One of the most crucial aspects of agriculture is soil quality, which is evaluated for pH, nutrients, and moisture content. The soil quality should be judged before commencing any farming operations. The nature and interaction of nanomaterials with

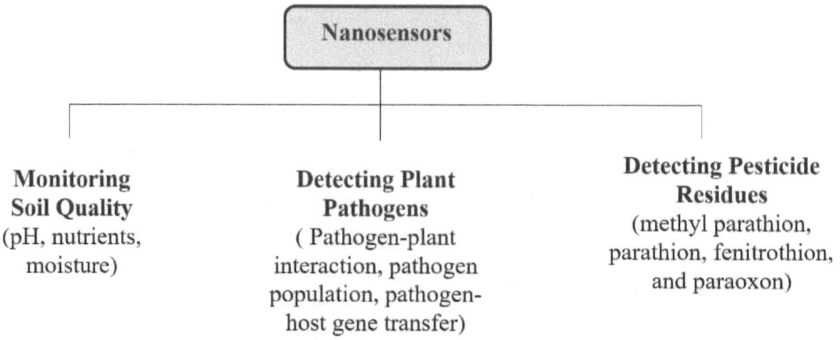

FIGURE 13.5 Classification of Nanosensors.

soil will determine their chemical, physical, and biological transformations. Particle aggregation will happen on its own when nanomaterials are put into environmental areas. Aggregation reduces surface area, which has an impact on the reactivity of the nanomaterials. There are two possible forms of aggregation: hetero-aggregation and homo-aggregation [25].

Contaminating pathogens were identified using nanobiosensors. The detection of agroterrorism agents was done using gold-nanoparticle-based nanobiosensors. Gold-particle-based nanoparticle-based sensors have been created to identify plant diseases. DNA is the bioproduct which is found in most of the nanobiosensors. Furthermore, pathogen–plant interactions, pathogen population genetics, and pathogen–host gene transfer can all be observed with the use of these nanobiosensors for the prediction of plant disease. The ailment known as witches broom disease in lemon plants is caused by Phytoplasma aurantifolia [26]. Nanobiosensors that exploit the quantum dots fluorescence resonance energy transfer phenomenon have been used to diagnose this condition. The most widely used type of nanobiosensors are metallic nanoparticle-based ones, which have a broad range of biorecognition components.

The pesticides are the chemicals which are used to reduce the amount of pests in an area, increasing agricultural yield by limiting damage. Different pesticides are typically administered at different times to plants as they grow. However, because of their toxicity, pesticide residue in agricultural products might pose a chemical risk to the agro-industry. Pesticides like methyl parathion, fenitrothion, and paraoxon [27] can be detected using a variety of detection methods, including biosensors that use surface Plasmon resonance, enzymatic sensors that use multiwalled carbon nanotubes or single-walled carbon nanotubes that are equipped with acetylcholinesterase, and others. A hollow fibre pencil-based graphite comprising multiwalled carbon nanotubes and CuO nanoparticles has been used to create such nanobiosensor [28].

13.1.4.13 Genetic Improvement

Increasing crop yields in a sustainable manner and minimizing overuse of fertilizers and pesticides is the main goal of current initiatives. Thus, it is imperative to use cutting-edge plant breeding technologies that have the capacity to create heritable targeted mutations in certain genomic regions, alter genes by knockout or knock in, and alter epigenetic alterations in order to address the shortcomings of traditional plant breeding techniques. Utilizing NPs as nanocarriers by forming a binding complex with bio-modifier molecules [29] and delivering them into plant cells is a powerful technique for the production of genetically modified organisms.

Depending on the basic material employed, NPs for gene delivery can be categorized as carbon-based, silicon-based, metallic, or polymer-based. The genetic contents delivered by each form of NP vary, accordingly, DNA and RNA can be carried by carbon nanotubes (CNTs) [30]. But the genetic cargo that the metallic NPs can carry is DNA [31]. Furthermore, polymeric NPs (such as polyethyleneimine and PEG) have the ability to introduce encapsulated RNA, DNA and proteins into cells while silicon-based nanoparticles [32] carry proteins and DNA.

13.1.4.14 Pest Management

Pesticides have been utilized extensively in agriculture and are regarded as one of the key elements of crop protection strategies. The traditional use of synthetic pesticides pollutes the environment by contaminating ground water. Researchers from all over the world began working towards the creation of a new class of pesticide, known as a "nanopesticide," based on the concepts of nanotechnology in order to address the issues related to the use of conventional pesticides.

The enhanced solubility of active ingredients, enhanced formulation stability, gradual release of active ingredient, and enhanced mobility due to reduced particle size and increased area are the main advantages. Nano formulations based on the chemical carriers are categorized as organic polymer-based formulations, lipid-based formulations, nanosized metals and metal oxides, clay-based nanomaterials, nano emulsions, nanosuspensions, etc. Pesticide formulations with active chemicals dispersed in a liquid media, such as crystalline or amorphous solid nanoparticles, are known as nanosuspensions, sometimes known as nanodispersions. A nanoemulsion is created when a chemical's active component is distributed in nanoscale droplets in water and surfactant molecules are trapped at the pesticide–water interface. A polymer-based delivery device [33] serves as a protective reservoir cover, promotes controlled pesticide release, and enhances the dispersion of active components in aqueous environments.

13.1.4.15 Food Industry

Food produced by applying nanotechnology to food processing, production, security, and packaging is referred to as nanofood. In the post-harvest food processing industry, nanotechnology holds enormous promise. It changes the particle size, size distribution, potential cluster formation, surface charge, and flavour, texture, and consistency of food, or masks an off-putting taste or odour. Animal flesh, fruits, veggies, milk products, fast-food items, baked goods and confectionery products can all use edible nano-coatings as moisture and gas shields. They also give manufactured goods flavour, colour, enzymes, antioxidants, anti-browning chemicals, and a longer shelf life. Food-borne illnesses can be prevented by producing healthier food with reduced fat, sugar, and salt content thanks to nanotechnology. TiO_2 and silicon dioxide (SiO_2) were reportedly approved for use as food additives.

Food quality loss is mainly caused by oxygen, water permeability, and ethylene, all of which are extremely sensitive to freshly harvested fruits and vegetables. Food packaging is therefore crucial to finding a solution to this problem. By acting as a barrier to stop oxygen and other gases that cause spoiling from scavenging, packing serves as a physical safeguard to protect food from temperature changes, external shock, vibration, and microbial infection. Its main goals are to suppress deterioration and contamination. The performance of the food packaging material is improved by the addition of active nanoparticles to the polymer matrices, which also give functional properties like scavenging, antioxidant, and antibacterial properties that extend the shelf life of the packaged food products.

13.1.5 Conclusion

Because of their distinct behaviour, nanoparticles are being used to improve quality of life in a variety of disciplines. The current widespread use of nanoproducts has increased life expectancy. The food sector and agriculture stand to gain greatly from the potential applications and advantages of nanotechnology, which could boost world food production. Precision farming and nano-priming are two examples of the cost-effective farming methods that farmers are developing with the aid of nanoproducts and nanomaterials. By serving as genetic cargo, nanotechnology has a role in genetic engineering as well. By tackling the main issues like poverty, malnutrition, and food scarcity, nanotechnology helps us to reach sustainable development goals. The development of nanoparticles and nanosensors has enormous promise for improving agriculture with proper precautions. Despite its numerous applications, nanotechnology is not accepted to the core extent. It has to emerge in every part of the world. Every new innovation has its impact on its own. The researchers have to work on the impacts of the nanotechnology on the environment to clear the challenges and fulfil the gap between the technology and the environment. The ability of nanotechnology to function within the confines of a sustainable agricultural system will ultimately determine its fate. The way the public feels about nanotechnology is also very important in influencing its future direction.

REFERENCES

1. Nadeem, B., Kammakakam, I., and Falath, W. 2021. Nanomaterials: A review of synthesis methods, properties, recent progress, and challenges. *Mater. Adv.* 2:1821–1871.
2. Lowry, G.V., Avellan, A., and Gilbertson L.M.. 2019. Opportunities and challenges for nanotechnology in the agri-tech revolution. *Nat. Nanotechnol.* 14:517–522.
3. Sajid, M., and Płotka-Wasylka, J. 2020. Nanoparticles: Synthesis, characteristics, and applications in analytical and other sciences. *Microchem J.* 154:104623.
4. Martin, Š., Illa, R., Ondřej, Z., Martin, Š., Pavel, D., and Marek, K. 2021. Effects of titanium dioxide nanomaterials on plants growth. *Nanotechnology in Plant Growth Promotion and Protection,* ed. Avinash P Ingle. Hoboken, NJ: Wiley & Sons Ltd.
5. Rana, A., Pathak, S., Kumar, K., Kumari, A., Chopra, S., Kumar, M., Kamil, D., Srivastava, R., Kim, S.-K., Verma, R., and Sharma, S.N. 2024. Multifaceted properties of TiO2 nanoparticles synthesized using *Mangifera indica* and *Azadirachta indica*

plant extracts: Antimicrobial, antioxidant, and non-linear optical activity investigation for sustainable agricultural applications. *Mater. Adv.* 5:2767.

6. Alisha, A S, A. and Thangapandiyan, S. 2019. Comparative bioassay of silver nanoparticles and malathion on infestation of red flour beetle, Tribolium castaneum. *J. Basic Appl. Zool.* 80:55.
7. Varghese, J., Rehaan Chandan, M., and Shanthakumar, S. 2020. Fixed bed column study for pesticide removal using silver nanoparticles-embedded polyurethane foam and glass beads. *Chem. Eng. Commun.* 207(10):1337–1346.
8. Mondal, A., Basu, R., Das, S., and Nandy, P. 2011. Beneficial role of carbon nanotubes on mustard plant growth: An agricultural prospect. *J. Nanopart. Res.* 13(10):4519–4528.
9. Serag, M.F., Kaji, N., Gaillard, C., Okamoto, Y., Terasaka, K., Jabasini, M., Tokeshi, M., Mizukami, H., Bianco, A., and Baba, Y. 2011. Trafficking and subcellular localization of multiwalled carbon nanotubes in plant cells. *ACS Nano.* 5:493–499.
10. Mejias, J., Salazar, F., Pérez Amaro, L., Hube, S., Rodriguez, M., and Alfaro, M. 2021. Nanofertilizers: A cutting-edge approach to increase nitrogen use efficiency in grasslands. *Front. Environ. Sci.* 9:52.
11. Rahale, S. 2011. Nutrient Release Pattern of Nanofertilizer Formulation. Ph.D. Thesis, Tamilnadu Agricultural University, Coimbatore, India.
12. El-Gazzar, N., Almaary, K., Ismail, A., and Polizzi, G. 2020. Influence of *Funneliformis mosseae* enhanced with titanium dioxide nanoparticles (TiO_2NPs) on *Phaseolus vulgaris* L. under salinity stress. *PLoS ONE.* 15:235355.
13. Thirugnanasambandan, T. 2021. Advances of engineered nanofertilizers for modern agriculture. In *Plant-Microbes-Engineered Nano-Particles (PM-ENPs)* Nexus in Agro-Ecosystems . Berlin/Heidelberg, Germany: Springer, pp. 131–152.
14. Jariwala, D., Sangwan, V.K., Lauhon, L.J., Marks, T.J., and Hersam, M.C. 2013. Carbon nanomaterials for electronics, optoelectronics, photovoltaics, and sensing. *Chem. Soc. Rev.* 42: 2824–2860.
15. Khan, M.Z.H., Islam, M.R., Nahar, N., Al-Mamun, M.R., Khan, M.A.S., and Matin, M.A. 2021. Synthesis and characterization of nanozeolite based composite fertilizer for sustainable release and use efficiency of nutrients. *Heliyon.* 7:6091.
16. Bryaskova, R., Pencheva, D., Nikolov, S., and Kantardjiev, T. 2011. Synthesis and comparative study on the antimicrobial activity of hybrid materials based on silver nanoparticles (AgNPs) stabilized by polyvinylpyrrolidone (PVP). *J. Chem. Biol.* 4:185–191.
17. Krishnaraj, C., Ramachandran, R., Mohan, K., and Kalaichelvan, P.T. 2012. Optimization for rapid synthesis of silver nanoparticles and its effect on phytopathogenic fungi. *Spectrochimica Acta A: Mol. Biomol. Spectrosc.* 93:95–99.
18. Reddy, G.B., Girija Mangatayaru, K., Madhusudan Reddy, D., Krishna, S.B.N., and Golla, N. 2022. *Biosynthesis and characterization methods of copper nanoparticles and their applications in the agricultural sector.* In: Abd Eslalam, K.A. (ed.), *NanoBiotechnology for Crop Protection.* Radarweg, Amsterdam: Elsevier, pp. 45–80
19. Viet, P.V., Nguyen, H.T., Cao, T.M., and Hieu L.V. 2016. *Fusarium* antifungal activities of copper nanoparticles synthesized by a chemical reduction method. *Journal of Nanomaterials.* 7:1957612.
20. Lu, B., Lv, X, and Le, Y. 2019. Chitosan-modified PLGA nanoparticles for control-released drug delivery. *Polymers.* 11:304.
21. Paulraj, G.M., Ignacimuthu, S., Rajiv, G.M., Shajahan, A., Ganesan, P., Packiam, S.M., and Al Dhabi, N.A. 2017. Comparative studies of Tripolyphosphate and

Glutaraldehyde cross-linked chitosan-botanical pesticide nanoparticles and their agricultural applications. *Int. J. Biol. Macromol.* 104:1813–1819.
22. Ghazali, S.A.I.S.M., Sarijo, S.H., and Hussein, M.Z. 2021. New synthesis of binate herbicide-interleaved anionic clay material: Synthesis, characterization and simultaneous controlled-release properties. *J. Porous Mater.* 28: 495–505.
23. Takeshita, V., Carvalho, L.B., Galhardi, J.A., Munhoz-Garcia, G.V., Pimpinato, R.F., Oliveira, H.C. et al. 2022. Development of a preemergent nanoherbicide: From efficiency evaluation to the assessment of environmental fate and risks to soil microorganisms. *ACS Nano*. 2:307–323.
24. Rojas, S., Rodríguez-Diéguez, A., and Horcajada, P. 2022. Metal–organic frameworks in agriculture. *ACS Appl. Mater. Interfaces* 14:16983–17007.
25. Ullah, N., Mansha, M., Khan, I., and Qurashi, A. 2018. Nanomaterial-based optical chemical sensors for the detection of heavy metals in water: Recent advances and challenges. *TrAC Trends in Analytical Chemistry*. 100:155–166.
26. J. Wu, F. 2021. *Accumulation and Impact of Metallic Nanomaterials in the Terrestrial Environment.* Leiden, Zuid Holland, Netherlands: Leiden University.
27. Duhan, J.S., Kumar, R., Kumar, N., Kaur, P., Nehra, K., and Duhan, S. 2017. Nanotechnology: The new perspective in precision agriculture, *Biotechnology Reports*. 15:11–23.
28. James, A., Franco, F., Merca, E.M., Rodriguez, S., Johny, F., Balidon-Veronica, P., Divina, M., Evangelyn, A., Alocilja, C., and Fernando, L.M. 2019. DNA based electrochemical nanobiosensor for the detection of Phytophthora palmivora causing black pod rot in cacao (Theobroma cacao l.) pods. *Physiol. Mol. Plant Pathol.* 107:14–20.
29. Gholivand, M.-B., Akbari, A., and Norouzi, L. 2018. Development of a novel hollow fiber-pencil graphite modified electrochemical sensor for the ultra-trace analysis of glyphosate. *Sensors and Actuators B: Chemical.* 272:415–424.
30. Demirer, G.S., Zhang, H., Matos, J.L., Goh, N.S., Cunningham, F.J., Sung, Y., et al. 2019. High aspect ratio nanomaterials enable delivery of functional genetic material without DNA integration in mature plants. *Nat. Nanotechnol.* 14:456–464.
31. Bates, K., and Kostarelos, K. 2013. Carbon nanotubes as vectors for gene therapy: Past achievements, present challenges and future goals. *Adv. Drug Deliv. Rev.* 65:2023–2033.
32. Zhao, X., Meng, Z., Wang, Y., Chen, W., Sun, C., Cui, B., et al. 2017. Pollen magnetofection for genetic modification with magnetic nanoparticles as gene carriers. *Nat. Plants.* 3:956–964.
33. Zhou, Y., Quan, G., Wu, Q., Zhang, X., Niu, B., Wu, B., et al. 2018. Mesoporous silica nanoparticles for drug and gene delivery. *Acta Pharmaceutica Sinica B.* 8:165–177.

14 Recent Advancements in Nanotechnology

A Comprehensive Review of Sustainable Agriculture

Rahul Joshi, Suman Kumari, and Krishna Pandey

14.1 INTRODUCTION

In the present digital age, the information communication technology (ICT) field is expanding rapidly, with the development of innovative and intelligent devices that can significantly enhance industrial and agricultural technologies. Agricultural issues have often impeded the country's progress. The only resolution to this predicament lies in smart agriculture, which optimizes conventional farming techniques. To update agricultural technology and precision farming, it is necessary to include advanced computer methods such as the Internet of Things (IoT), Artificial Intelligence (AI), big data, robotic systems, unmanned ground vehicles (UGV), unmanned aerial vehicles (UAV), and so on [1][2]. Advanced agricultural technology has decreased the influence of weather conditions, fertilizer usage, and water on farming. Simultaneously, it has improved worker safety, efficiency, and output rates. Furthermore, it aids farmers in effectively managing and regulating production risks while enhancing the capacity to anticipate process results. IoT-based smart farming is a system designed to oversee the agricultural field using sensors such as light, humidity, temperature, soil moisture, and crop health and to automate field activities. Data collected from IoT devices is analyzed using AI-based models to monitor crops, survey and map fields, detect abiotic/biotic stress, and forecast environmental conditions and agricultural output.

Smart farming focuses on implementing digital agriculture via artificial intelligence, the Internet of Things, remote sensing, robots, cloud computing, and big data analytics. These cutting-edge agricultural techniques decrease costs and enhance crops' quality, value, and productivity. Additionally, it aids in maintaining equilibrium in the supply and demand chains of agricultural goods within the marketplace. Hence, intelligent digital farming can enhance agriculture's sustainability by enabling real-time data gathering, control, and decision-making across agricultural operations [3]. Decision-making in the embedded system may use sensor data as input and immediately transmit the outcomes to the machine for precise management actions.

Farmers may use cloud-based farm management applications to store and access information about the market, field data, newly developed agricultural equipment, weather, and climatic conditions [4]. This database facilitates prompt management decision-making for framers. Cloud services enable the resolution of farmers'

DOI: 10.1201/9781003487005-14

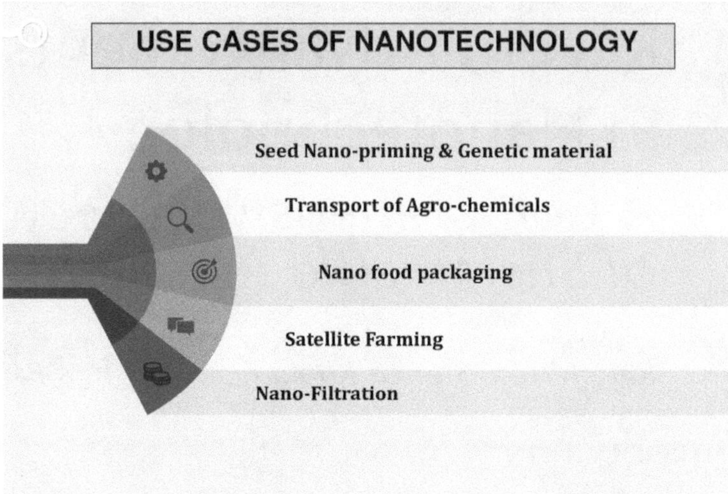

FIGURE 14.1 Diverse applications of nanotechnology in agricultural sector.

inquiries by offering fast answers rooted in specialized knowledge. They can get professional guidance via online cloud-based databases. The farm management system is a web-based application that provides virtual instruction to farmers, imparting knowledge about illnesses, pesticides, and agricultural maintenance [5].

The agriculture industry may undergo a revolutionary transformation via big data, which involves a cloud-based ecosystem equipped with appropriate tools and software to combine diverse data sources seamlessly. The potential for using big data in the current agricultural sector is substantial. The capacity to monitor tangible objects, collect up-to-the-minute information, and predict occurrences may significantly transform digital agriculture. Data analytics revolutionize the agricultural industry by addressing several challenges, such as feeding a rising population, ensuring ethical use of pesticides, enhancing farm equipment efficiency, and handling supply chain complexities. These analytics provide valuable insights to assist decision-making processes in the sector. Figure 14.1 displays the numerous use cases for nanotechnology.

14.2 NANOTECHNOLOGY: A NEW ERA IN FARMING

Precision agriculture that utilizes nanotechnology involves using computers, global positioning systems (GPS), and remote sensing devices to accurately assess crop-related and environmental factors [6]. Precision agriculture methods use efficient monitoring tools and protocols to reduce the need for pesticides, fertilizers, and herbicides. This approach entails the deliberate and regulated dispensation of agrochemicals onto specific targets to optimize nutrition absorption and enhance resistance against diseases. These include Nanoscale transporters, Nanosensors, Nanofertilizers (NFs), Nanoherbicides, and Nanopesticides. This advanced technology may assist farmers in making informed choices on utilizing fertilizers, insecticides, and other resources. The use of smartphone-based biosensors with cloud computing may enhance the process of real-time monitoring and decision-making. The assessment also emphasizes the potential of nanotechnology in diminishing agro-waste and producing valuable goods, which may substantially affect agriculture's sustainability and profitability. Nevertheless, the assessment highlights apprehensions about implementing precision agriculture in underdeveloped nations, including data administration, ownership, privacy, infrastructure, and socio-economic factors. It is essential to address these challenges to implement precision agriculture in India and other emerging nations effectively. Figure 14.2 shows the significant milestones achieved in nanotechnology.

14.3 UTILIZING NANOPARTICLES FOR SUSTAINABLE AGRICULTURE

Nanoparticles possess distinct physical, chemical, and biological characteristics compared to their larger counterparts due to their diminutive dimensions and elevated surface area-to-volume ratio. Using nanoparticles to enhance seed germination, plant development, nutritional provision, and stress mitigation might substantially augment crop yields and diminish production expenses. Nanoparticles have exceptional versatility and may be used across various industries, such as medicine, electronics, energy, and environmental remediation, because of their unique features.

14.3.1 A New Form of Carbon: Nanotubes

A unique carbon structure may be conceptualized as a cylindrical tube created by rolling a two-dimensional graphene sheet. There are two main types of nanotubes: "single-walled nanotubes (SWNTs) and multi-walled nanotubes (MWNTs). The material's tensile strength, which measures its ability to withstand stretching forces, is 200 GPa. This makes it very suitable for use in reinforced composites and nanoelectromechanical systems. Furthermore, it may exhibit metallic or semiconducting properties, which presents significant prospects for advancing electrical circuits and even whole nanodevices. "The nanotube systems consist of tightly coiled graphitic layers arranged in cylindrical forms. Fluorescent nanoparticles or quantum dots have been developed in recent times for labelling plant proteins" [7].

14.3.2 A Tiny Semiconductor: Quantum Dots

Semiconductor quantum dots typically exhibit high quantum yield and molar extinction coefficients. They have broad absorption and narrow, symmetric fluorescence spectra covering the ultraviolet to near-infrared range. Furthermore, QDs possess significant effective excitation, high photobleaching resistance, and exceptional photochemical degradation resistance. These materials exhibit remarkable fluorescence and quantum confinement of charge carriers, and they can adjust the band energy based on their size [8]. They have unique spectral properties in comparison to traditional organic dyes. Consequently, they have emerged as a novel class of

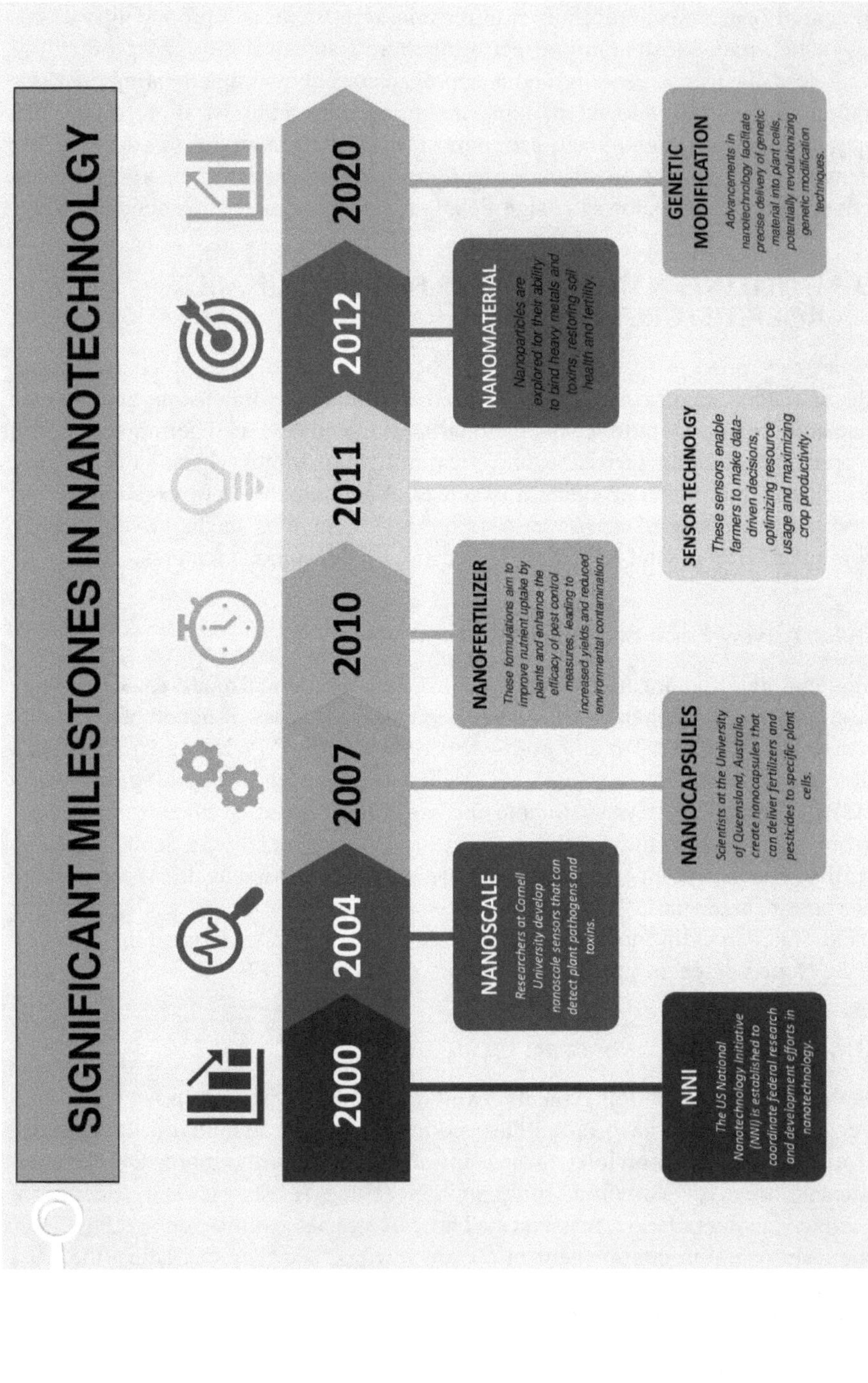

FIGURE 14.2 Timeline outlining significant milestones in agriculture.

fluorophores for applications in bioimaging and biosensing [9]. Quantum dots also serve as photocatalysts for the light-induced chemical transformation of water into hydrogen, which may be used to produce solar fuel.

14.3.3 A Multinational Plasmonic: Nanorods

Multimodal plasmonic materials, capable of effectively coupling sensing phenomena and regulating energy at adjustable sizes, may be linked to MEMS and elicit targeted field responses [10]. The gold nanorods substantially impacted the physiological changes in the watermelon plant and demonstrated phytotoxicity toward the plants, especially at high concentrations.

14.3.4 A New Era of Techniques: Micro and Nano Encapsulation

Encapsulation refers to enveloping an item with a protective layer or embedding it inside a uniform or diverse matrix. Consequently, this procedure yields capsules that possess several advantageous characteristics [11]. Encapsulation technologies provide many advantages, including safeguarding chemicals or objects from unfavorable environments, precise targeting, and regulated release [12]. Various encapsulation strategies are used based on the size and form of the capsules. Macro encapsulation/coating produces capsules on a larger scale, whereas micro- and nanoencapsulation results in particles of smaller sizes, namely in the micro and nanoscale range. Nanocapsules are vesicular structures where chemicals are contained inside a cavity of an inner liquid core surrounded by a polymeric membrane.

14.3.5 Colloidal Dispersions of Nanoscale: Nanoemulsions

Nanoemulsions consist of tiny droplets in an oil/water mixture with diameters below 100 nm [13]. While there may not be any fundamental differences between nanoemulsions and microemulsions, the physical characteristics of nanoemulsions might vary significantly from those of microscale emulsions. Nanoemulsions exhibit much greater surface area to volume ratio, Laplace pressure, and elastic modulus than regular emulsions, owing to the smaller droplet size.

14.4 COMPARISON BETWEEN TRADITIONAL FERTILIZERS AND NANOFERTILIZERS

Farmers or producers use traditional fertilizers by applying them to the soil via surface broadcasting, subsurface placement, fertigation, or irrigation water. Nevertheless, a significant proportion of the fertilizers are released into the atmosphere or water bodies, contaminating our ecosystems. For instance, 75% of the urea's nitrogen content is lost by volatilization (as NH_3) or emission as N_2O or NO or through NO3 leaching or runoff into water bodies after it is applied in the field. The present nitrogen fertilizers encounter the issue of poor nitrogen utilization efficiency, with less than 20% of nitrogen being effectively used. Consequently, the loss of nitrogen in the environment leads to eutrophication and a rise in greenhouse gas emissions. Reports indicate that the soil loses a substantial number of essential macronutrients, including nitrogen (N), phosphorus (P), and potassium (K), at varying rates. These rates range from 40–70% for nitrogen, 80–90% for phosphorus, and 50–90% for potassium. This leads to a substantial depletion of resources. The surplus phosphorus undergoes a process called "fixation" in the soil, where it forms chemical links with other nutrients and becomes inaccessible for plants to absorb. The key advantages of nano fertilizers compared to traditional chemical fertilizers lie in their capacity to manage the release of nutrients in crops via a delayed and regulated delivery method. The sluggish delivery of nutrients is linked to the encapsulation or immobilization of nutrients using nanomaterials. Growers may enhance crop development by exploiting the gradual supply of nutrients, which ensures a steady and prolonged release of nutrition to plants.

Furthermore, nanofertilizers need minimal quantities, decreasing transportation and field application expenses. A notable benefit is the ability to minimize salt formation in soil due to its need in tiny proportions. An additional advantage of using nanofertilizers is their ability to be synthesized by the specific nutritional needs of intended crops. Biosensors may be affixed to a novel fertilizer that regulates the release of nutrients based on the soil's nutrient levels, crop development stage, or environmental circumstances.

14.5 SUSTAINABLE CROP DEVELOPMENT WITH PRECISION FARMING

Precision agriculture is a farming method that uses modern technologies and data-driven tools to enhance crop yields and efficiently manage resources. The goal is to minimize waste, improve efficiency, and boost profitability. Nanotechnology is a scientific and technological subject focusing on materials and structures at a tiny level. Nanotechnology has the potential to transform precision agriculture by providing farmers and producers with innovative tools and methods to improve crop yield. Nanotechnology and precision agriculture have distinct focuses but also exhibit interconnected characteristics. Nanotechnology has the potential to develop novel materials and instruments that enhance precision agricultural techniques. Nanosensors can monitor soil and plant health, enabling precise and efficient crop management. Nanoparticles may improve the transportation of nutrients and insecticides to plants, minimizing loss and boosting efficiency.

Nanotechnology enhances agricultural yields, decreases waste, and lowers environmental consequences in precision agriculture. Nanoscale alteration of materials provides distinct advantages and benefits compared to traditional farming methods. Nanosensors can provide real-time soil and plant health monitoring, assisting farmers in optimizing crop management via precise water and fertilizer requirements data. Nanoparticles may serve as carriers for insecticides and fertilizers, hence decreasing their ecological footprint [14]. Nanotechnology may improve agricultural resources' characteristics, such as plant fibers and seeds, by increasing their resistance to pests and weathering. Figure 14.3 demonstrates the substantial capabilities of nanotechnology.

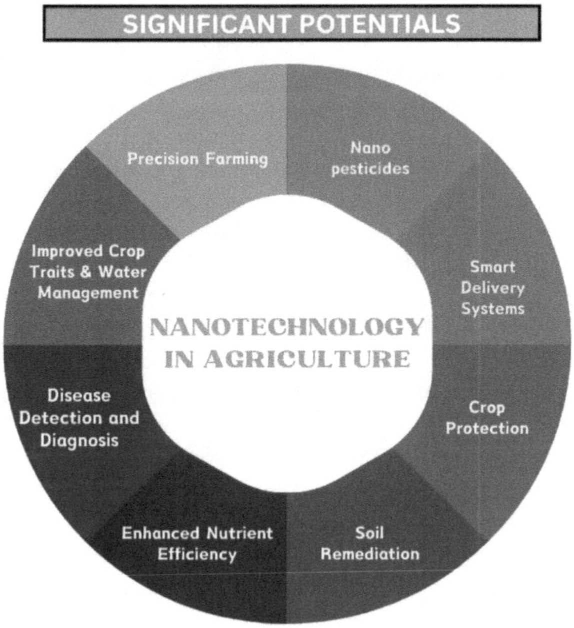

FIGURE 14.3 Significant potentials of nanotechnology in agriculture.

The following points explain the significant potential of nanotechnology in precision farming:

1. ***Enhanced absorption of nutrients:*** The growing need for higher food production worldwide requires new agricultural methods to improve crop yield and reduce environmental harm. Using nano-scale materials, namely NFs, has enhanced crop nutrient utilization efficiency and minimized the negative impacts of over-fertilization [15]. Nanotechnology may significantly improve crop fertilizer delivery by enclosing nutrients in nanoparticles, enabling precise and regulated release. Nutrient fertilizers are usually created by enclosing nutrients in nanoscale carriers, such as metal oxide nanoparticles or polymeric nanocapsules. This method allows for the nutrients to be released in a controlled manner, reducing losses caused by leaching or volatilization.
2. ***Intelligent Environmental Surveillance:*** Nanotechnology is a powerful tool for enhanced environmental monitoring for precision agriculture. The primary use is nanosensors that continuously monitor soil, water, and plant factors. Nanosensors provide crucial data for improving agricultural management techniques. They can detect variations in soil characteristics such as

moisture, nutrient levels, and pH. An excellent example of this use is zinc oxide nanoparticles as nanosensors, which are very skilled at detecting phosphorus levels in the soil [16]. The accurate detecting capabilities of these nanosensors allows for the best use of water and fertilizers, avoiding over- or under-fertilization and promoting better crop development.

3. ***Better Pest Management:*** Combining nanopesticides with precision agriculture improves pest management by adjusting pesticide use according to live monitoring of insect levels and environmental factors [17]. Nanoparticles may be designed to target specific pests or plant components, providing precise delivery of pesticides and the creation of novel, potent insecticides that work well in smaller amounts. Nanotechnology-based PA decreases environmental effects by enabling accurate pesticide delivery, reducing pesticide discharge into the environment. Examples include using nanoparticles to transport RNA interference (RNAi) molecules for precise and pest efficient management and employing nanocapsules for specific pesticide delivery to enhance effectiveness and minimize environmental impact. Nanotechnology-based precision agriculture enhances safety for farmers and consumers by reducing direct pesticide exposure by creating less harmful pesticides. Nanobiosensors may help identify pests and illnesses early, decreasing the need for widespread pesticide use and enhancing safety for producers and consumers.
4. ***Analytical data mining:*** Data analytics is crucial in precision agriculture, enabling farmers to make well-informed choices on planting, fertilizing, and harvesting crops. Utilizing NPs may enhance the accuracy and precision of data collecting, resulting in more dependable analytics. For example, farmers might use NPs to improve their comprehension of their crops' condition by analyzing soil or water samples for specific chemicals or illnesses. Nanoparticles may function as sensors to detect pollutants in water and soil. Gold nanoparticles may be used for detecting heavy metals in soil samples.
5. ***Utilization of Nanomaterials in Enhancing Plant Development:*** Nanomaterials are being increasingly used in agriculture to improve plant growth and production. Nanocarbon, nanocellulose, and Nanolignocellulose are nanomaterials that have shown potential impacts on plant development. Nanocarbons such as carbon nanotubes, graphene, and fullerenes have shown significant promise in enhancing plant development. Research shows that applying carbon nanotubes to tomato plants led to a substantial boost in growth [18]. Graphene oxide (GO) treatment may also improve plant growth.
6. ***Variable Rate Technology:*** VRT uses data and automation to adjust the application of fertilizer, crop protection agents, seeds, and irrigation water at varying rates in various areas without the need for human adjustment by the producer. Variable rate technology in precision agriculture involves the precise application of herbicides, insecticides, lime, gypsum, crop nutrients, and sowing according to the specific needs of the crops, as well as the detection of weeds and damaged crops. VRT uses GPS and GIS technology to pinpoint specific locations in the field for material application.

7. ***Autonomous Equipment:*** Incorporating nanosensors into automated machines may improve the accuracy and effectiveness of agricultural tasks. Nanosensors are minuscule sensors created to identify specific substances or environmental circumstances. Nanosensors may enhance precision and efficacy in automated machines via numerous means. "Nanosensors can monitor soil moisture levels, enabling mechanical irrigation systems to change water supply rates in real time. Nanosensors can provide precise and timely feedback on soil moisture levels to minimize overwatering and underwatering, which may harm crop development" [19].
8. ***Genetic engineering:*** Nanotechnology may be integrated to enhance the efficiency of crop enhancement and plant transformation. Genetic engineering in agriculture has led to significant advancements, such as creating high-yield and nutritionally rich crops and crops resistant to herbicides. This also resulted in crops resisting biotic pressures such as insects and diseases and abiotic stresses like drought and heat. Despite yielding unexpected outcomes, genetic engineering faced constraints because of the absence of diverse and high-throughput methods for introducing biomolecules into plant cells [20].

14.6 BENEFITS OF NANOTECHNOLOGY IN FARMING ECOSYSTEMS

1. ***Enhanced Seed Germination and Plant Development:*** Increased agricultural yield requires a greater seed germination rate. Environmental pollution and several abiotic stresses harm seed germination and seedling health. Reduced seed viability is a notable concern in dry and semi-dry areas due to the impact of non-living factors on the timing of seed germination. Laboratory-tested seeds with high germination rates sometimes need to perform better in field experiments [21]. These issues need a well-planned approach to solving them. Multiple techniques may be used to enhance a poor seed germination rate. Priming seeds using nanoparticles has recently piqued the scientific community's attention. Recently, the impact of nanomaterials on seed germination to enhance the germination rate has been examined closely.
2. ***Alleviating Plant Stress from Living and Non-living Factors:*** Biotic stress is produced by living creatures like pests and diseases, whereas abiotic stress is caused by non-living causes such as drought, salt, and heavy metal toxicity. NPs may help mitigate abiotic stress by decreasing oxidative harm and boosting plant antioxidant defense mechanisms. Various categories of NPs, including metal-based, metal oxide-based, carbon-based, and polymer-based, have been studied for their ability to reduce plant stress. Soil application of silver nanoparticles improved plant development and decreased biotic stress caused by Aspergillus in rice by hindering pathogen growth. Zinc oxide nanoparticle sprays helped reduce oxidative damage in tomato plants under abiotic stress conditions [22].
3. ***Enhanced Micronutrient Provision:*** Plants need trace levels of micronutrients for their growth. On the contrary, most agricultural land lacks many

micronutrients. Crops in these places have reduced yields because of a need for essential micronutrients. Consuming meals lacking in micronutrients may impact human health by producing anemia, stunted development, impaired reproductive functions, and lower mental and physical capacities. The Food and Drug Administration (FDA) in Silver Spring, Maryland, USA, authorizes nutritious food to promote a healthy and long life [23]. These soils need supplementation in the form of fertilizers for the proportion of micronutrients applied as fertilizer that reaches the plant. Surplus micronutrients are carried away by rain into nearby water sources. Furthermore, micronutrients in the soil slowly bind with other compounds and are no longer accessible to the plant.

4. ***Enhanced Plant Fertilization at Lower Dosage:*** NF has distinct physical and chemical characteristics compared to bulk materials. Nanorock phosphate, when sprayed, enhances phosphorus availability in plants by preventing fixation by soil due to its nanoform. Nanofertilizers are receiving significant focus in the agricultural scientific community because of their excellent penetration capability, compact size, and large surface area. This material has distinctive features that differentiate it from similar bulk materials. NF is either produced by chemical synthesis or extracted from plants utilizing nanotechnology. The production technique boosts its ability to promote soil fertility and increase crop yield. Nitrogen fixation may help photosynthesis by enhancing crop production and quality via efficient nutrient absorption and minimizing fertilizer loss. They may regulate nutrient availability to align with crop development and sustain nutrient supply over the plant's growth cycle. Nanofertilizers may enhance soil fertility, are non-toxic, and are cost-effective due to their lower application rates. Creating nanocomposites might help meet the need for critical nutrients by using an intelligent delivery mechanism.
5. ***Reducing Chemical Dosage:*** The present surge in the global population has caused a sharp increase in food needs, resulting in remarkable growth in the global pesticide industry. Regrettably, several agrochemicals are entering the human food chain, damaging human and animal health, agriculture, and ecology. Increased pesticide application rates may lead to insect resistance, necessitating greater pesticide dosages. Pesticides have unintentionally significantly decreased the population of certain insects, including honey bees. Regrettably, these compounds are not exclusive to agricultural regions and may be found in the air, water, and soil, leading to environmental contamination [24]. Minimizing pesticide usage is essential for alleviating ecological degradation and reducing crop production expenses.

14.7 UTILIZING NANOTECHNOLOGY FOR THE LABELING, MONITORING, AND SURVEILLANCE OF AGRO PRODUCTS

Nanotechnology, which can control materials at the atomic and molecular scale, has great potential to transform other sectors, such as agriculture. Nanotechnology provides advanced methods and tools for tagging, tracking, and monitoring agricultural goods to guarantee safety, quality, and traceability throughout the supply chain.

1. ***Nanoparticle-based Labels:*** Nanotechnology is often used in agro-product management to create labels based on nanoparticles. These labels, sometimes including quantum dots or other nanomaterials, act as distinct markers for monitoring items from production to consumption. Quantum dots provide outstanding fluorescence characteristics that allow straightforward detection and monitoring, improving product traceability [25].
2. ***Nano barcodes:*** Nanostructured barcodes are an advanced use of nanotechnology in labeling agricultural products. Nanoscale barcodes created by sophisticated lithography processes serve as detailed identifiers for specific items. Nano barcodes encode crucial details like source, manufacturing date, and lot number to streamline tracking and verification throughout the supply chain [26].
3. ***Nanosensors, Nanofibers and Nanocoatings:*** Nanotechnology enables the creation of susceptible sensors that detect pollutants, infections, and rotting indications in agricultural goods. Nanomaterial sensors have unparalleled sensitivity, allowing for the early identification of possible dangers. The use of nanofibers and nanocoatings in agricultural product packaging is a notable progress made possible by nanotechnology. Nanofiber-based materials have outstanding barrier qualities that significantly increase the shelf life of perishable items [27]. Nanocoatings applied to surfaces protect against microbiological contamination and degradation, maintaining product integrity throughout storage and transit.
4. ***Nanoparticle-based Imaging:*** Nanoparticle imaging methods provide a non-invasive way to assess the quality and condition of agricultural products. Researchers may use nanoparticles with unique imaging characteristics, such as magnetic resonance or fluorescence, to observe interior structures and identify abnormalities [28].
5. ***Nanofluidic Devices:*** Nanofluidic devices are essential for doing on-site testing and analysis of agricultural goods. These little gadgets, including tiny channels and sensors at the nanoscale level, allow for quick and precise identification of impurities and harmful microorganisms.
6. ***Nanomaterial-based Filters:*** Nanotechnology provides advanced options for purifying water in agriculture by creating filters using nanomaterials. These filters, which include designed nanoparticles, efficiently eliminate impurities and toxins from irrigation water, guaranteeing the purity and safety of agricultural materials [29].
7. ***Nanoparticle-based Tracers:*** Nanoparticles are valuable tracers for monitoring the transport of agrochemicals in soil and plants. Researchers may track the distribution and longevity of pesticides, fertilizers, and other agricultural products by tagging them with nanoparticles.

8. ***Nanorobots, Nano-enabled Data Management Systems:*** Nanorobots, while still in the first phases of development, have significant promise for the precise delivery of nutrients, insecticides, and growth regulators in agriculture. These little robots, controlled by advanced technology, can accurately deliver agricultural substances at the cellular level. Nanorobots provide exceptional accuracy and effectiveness in applying agrochemicals, reducing waste and environmental harm. Nanotechnology-enhanced advanced data management systems smoothly integrate information collected by different sensors and surveillance devices. These devices use nanoscale computing and networking features to facilitate real-time analysis of data decision-making [30].

14.8 ANALYSIS OF ECONOMIC, LEGAL, AND SOCIAL ISSUES, AND HAZARDS IN NANOTECHNOLOGY

Nanotechnology in agriculture has the potential to transform the business by boosting crop output, strengthening soil health, and decreasing the environmental effects of agricultural methods. Despite its potential advantages, several economic, legal, social, and risks must be thoroughly examined and resolved.

14.8.1 Economical Issues in Nanotechnology

1. ***Cost of Technology, Market Accessibility and Dependency on Corporations:*** Implementing nanotechnology in agriculture often requires substantial upfront investment in research, development, and application. This might provide difficulties for small-scale farmers or underdeveloped nations with little resources [31]. Disparities in access to nanotechnology-based agricultural goods or services may result in inequities between big retail farms and smallholders. Integrating nanotechnology in agriculture might result in heightened reliance on significant companies for patented nano-enhanced goods or seeds, thus restricting options for farmers and strengthening corporate dominance over the agricultural sector.

14.8.2 Legal Issues in Nanotechnology

1. ***Regulatory Frameworks, Intellectual Property Rights (IPR) and Environmental Regulations:*** The regulations for nanotechnology in agriculture are continuously developing. There may need to be more than existing rules to handle the threats posed by nanoscale particles in farming [32]. There is the potential for developing problems with patents and intellectual property rights, especially regarding the ownership and management of nanotechnology-enabled agricultural technologies. This could stifle innovation and restrict access to helpful technology. Nanoparticles employed in agriculture might have unforeseen effects on the environment.

14.8.3 Social Issues in Nanotechnology

1. ***Acceptance and Perception, Impact on Farming Communities and Food Security:*** The public's perspective on nanotechnology in agriculture may differ, with some people expressing worries about its safety, ethical implications, and possible consequences that are unknown. Agricultural communities may have socio-economic repercussions due to the use of nanotechnology. These repercussions may include shifts in employment patterns, opportunities for rural development, and traditional farming techniques. It is of the utmost importance to consider the fair distribution of benefits and the participation of local communities in the decision-making processes. There are worries about the long-term consequences of nanotechnology for food safety, biodiversity, and sustainability. Although nanotechnology has the potential to raise agricultural yields and promote food security, there are also questions over its possible downsides.

14.8.4 Hazards Issues in Nanotechnology

1. ***Environmental Risks, Health, and Safety Risks and Ethical Hazards:*** Nanoscale materials employed in agriculture might have unforeseen environmental implications, such as soil and water pollution, as well as effects on creatures that are not the intended targets. To guarantee the long-term viability of the ecosystem, it is essential to evaluate and eliminate these threats. For farmers, farmworkers, consumers, and the public, there are concerns over the possible health and safety issues connected with exposure to nanoscale materials. These risks include threats to their health and safety. When minimizing these dangers, proper risk assessment and safety measures are essential. It is necessary to give serious thought to ethical issues to guarantee responsible and sustainable growth. Some examples of these issues include fair access to the advantages of nanotechnology, the possibility of misusing agricultural technologies enabled by nanotechnology, and the influence on traditional farming techniques.

14.9 CONCLUSION AND IMPLICATIONS OF THE STUDY

Despite being twenty years old, the advancements in nanotechnology science have been remarkable. The distinctive properties of nanoparticles make them valuable for improving the quality of life across several disciplines. Nanotechnology helps accomplish sustainable development objectives by tackling critical issues such as poverty, hunger, and food shortages in developing nations. It has yet to reach every part of the world. The adverse effects of nanoparticles accumulating in the environment and generating ecotoxicity cannot be ignored. Nanofertilizers provide economic advantages by decreasing leaching and volatilization compared to traditional fertilizers. The method can significantly boost farmers' profit margins by improving production and product quality. Although nanofertilizers have shown promising agricultural results, their marketability has yet to be a primary emphasis. Before using nanofertilizers on a large scale, it is crucial to thoroughly investigate the uncertainty around how nanomaterials interact with the environment and their possible impact on human health. Additional study is required to pinpoint the bottlenecks, minimize the negative effects, and optimize the advantages.

Agricultural experts often publish updated advice and technology advancements in agrarian techniques in local publications to assist farmers. The acceptance rate of such proposals or technology among rural communities could be more precise. Thus, governmental policy changes are necessary to finance research projects on farmers embracing new technology. The physical and chemical characteristics of soil particles determine the impact of nanoparticles on soil. Additional study is required to investigate how metal nanoparticles affect the terrestrial ecosystem, the interplay of contaminants in different climatic situations, and their impact on the rhizosphere while considering their characteristics in diverse soil types and plant species.

REFERENCES

1. Misra, Nandan Nitish, Yash Dixit, Ahmad Al-Mallahi, Manreet Singh Bhullar, Rohit Upadhyay, and Alex Martynenko. "IoT, big data, and artificial intelligence in agriculture and food industry." *IEEE Internet of Things Journal* 9, no. 9 (2020): 6305–6324.
2. Joshi, Rahul, Krishna Pandey, and Suman Kumari. "A metaverse: Significance and dimensions in various sectors." In *2024 11th International Conference on Computing for Sustainable Global Development (INDIACom)*, pp. 1781–1786. IEEE, 2024, New Jersey, USA.
3. Mehta, Chetram, Raghu, Nakul, Singh, Chandel, and Yash. Anil Rajwade. "Smart farm mechanization for sustainable Indian agriculture." (2020).
4. Ozdogan, Burak, Anil Gacar, and Huseyin Aktas. "Digital agriculture practices in the context of agriculture 4.0." *Journal of Economics Finance and Accounting* 4, no. 2 (2017): 186–193.
5. Ault, Aaron C., James V. Krogmeier, and Dennis Buckmaster. "Mobile, cloud-based farm management: a case study with trello on my farm." In *2013 Kansas City, Missouri, July 21–July 24, 2013*, p. 1. American Society of Agricultural and Biological Engineers, 2013, Michigan, USA.
6. Kaushal, Manoj, and Suhas P. Wani. "Nanosensors: frontiers in precision agriculture." *Nanotechnology: An Agricultural Paradigm* (2017): pp. 279–291. Singapore: Springer Nature.
7. Chahine, Nadeen O., Nicole M. Collette, Cynthia B. Thomas, Damian C. Genetos, and Gabriela G. Loots. "Nanocomposite scaffold for chondrocyte growth and cartilage tissue engineering: effects of carbon nanotube surface functionalization." *Tissue Engineering Part A* 20, no. 17–18 (2014): 2305–2315.
8. Androvitsaneas, Petros, Andrew B. Young, Christian Schneider, Sebastian Maier, Martin Kamp, Sven Höfling, Sebastian Knauer et al. "Charged quantum dot micropillar system for deterministic light-matter interactions." *Physical Review B* 93, no. 24 (2016): 241409.
9. Bakalova, Rumiana, Hideki Ohba, Zhivko Zhelev, Mitsuru Ishikawa, and Yoshinobu Baba. "Quantum dots as photosensitizers?" *Nature biotechnology* 22, no. 11 (2004): 1360–1361.

10. Noori, Rubia, Razi Ahmad, and Meryam Sardar. "Nanobiosensor in health sector: the milestones achieved and future prospects." *Nanobiosensors for Agricultural, Medical and Environmental Applications* (2020): pp. 63–90. Singapore: Springer Nature.
11. Rodríguez, Julia, María J. Martín, María A. Ruiz, and Beatriz Clares. "Current encapsulation strategies for bioactive oils: From alimentary to pharmaceutical perspectives." *Food Research International* 83 (2016): 41–59.
12. Ozdemir, Murat, and Tansel Kemerli. "Innovative applications of micro and nanoencapsulation in food packaging." *Encapsulation and Controlled Release Technologies in Food Systems* (2016): 333–378. https://onlinelibrary.wiley.com/doi/10.1002/9781118946893.ch12
13. Anton, Nicolas, and Thierry F. Vandamme. "Nano-emulsions and micro-emulsions: clarifications of the critical differences." *Pharmaceutical research* 28 (2011): 978–985.
14. Shang, Yifen, Md Kamrul Hasan, Golam Jalal Ahammed, Mengqi Li, Hanqin Yin, and Jie Zhou. "Applications of nanotechnology in plant growth and crop protection: a review." *Molecules* 24, no. 14 (2019): 2558.
15. Salem, Salem S., and Alsayed E. Mekky. "Biogenic Nanomaterials: Synthesis, Characterization, and Applications." In *Biogenic Nanomaterials for Environmental Sustainability: Principles, Practices, and Opportunities*, pp. 13–43. Cham: Springer International Publishing, 2024.
16. Mahmoud, Alaa El Din, and Manal Fawzy. "Nanosensors and nanobiosensors for monitoring the environmental pollutants." *Waste Recycling Technologies for Nanomaterials Manufacturing* (2021): 229–246.
17. Pradhan, Saheli, and Damodhara Rao Mailapalli. "Nanopesticides for pest control." *Sustainable Agriculture Reviews 40* (2020): 43–74.
18. Khodakovskaya, Mariya V., Bong-Soo Kim, Jong Nam Kim, Mohammad Alimohammadi, Enkeleda Dervishi, Thikra Mustafa, and Carl E. Cernigla. "Carbon nanotubes as plant growth regulators: Effects on tomato growth, reproductive system, and soil microbial community." *Small* 9, no. 1 (2013): 115–123.
19. Mallinger, Kevin, and Ricardo Baeza-Yates. "Responsible AI in Farming: A Multi-Criteria Framework for Sustainable Technology Design." *Applied Sciences* 14, no. 1 (2024): 437.
20. Abd-Elsalam, Kamel A. "Microbial Nanotechnology." *Microorganisms* 12, no. 2 (2024): 352.
21. Batty, Alex Lan, Kelvin Wilson Dixon, Mark Brundrett, and Krishnapillai Sivasithamparam. "Constraints to symbiotic germination of terrestrial orchid seed in a Mediterranean bushland." *New Phytologist* 152, no. 3 (2001): 511–520.
22. Zheng, Lei, Fashui Hong, Shipeng Lu, and Chao Liu. "Effect of nano-TiO 2 on strength of naturally aged seeds and growth of spinach." *Biological Trace Element Research* 104 (2005): 83–91.
23. Kumar, Amit, Indrakant K. Singh, Rashmi Mishra, Akanksha Singh, Naleeni Ramawat, and Archana Singh. "The role of zinc oxide nanoparticles in plants: A critical appraisal." *Nanomaterial Biointeractions at the Cellular, Organismal and System Levels* (2021): 249–267.
24. Wu, Charles, Su-Lin Lee, Cassandra Taylor, Jing Li, Yen-Ming Chan, Rajiv Agarwal, Robert Temple, Douglas Throckmorton, and Katherine Tyner. "Scientific and regulatory approach to botanical drug development: A US FDA perspective." *Journal of Natural Products* 83, no. 2 (2020): 552–562.

25. Rajmohan, Kumar Suresh, Ramya Chandrasekaran, and Sunita Varjani. "A review on occurrence of pesticides in environment and current technologies for their remediation and management." *Indian Journal of Microbiology* 60, no. 2 (2020): 125–138.
26. Mittal, Deepti, Gurjeet Kaur, Parul Singh, Karmveer Yadav, and Syed Azmal Ali. "Nanoparticle-based sustainable agriculture and food science: Recent advances and future outlook." *Frontiers in Nanotechnology* 2 (2020): 579954.
27. Cheng, Huai Ni, Kalson Teno Klasson, Tetsuo Asakura, and Qinglin Wu. "Nanotechnology in agriculture." In *Nanotechnology: Delivering on the Promise Volume 2*, pp. 233–242. American Chemical Society, 2016. Washington, USA.
28. Mohammad, Zahra H., and Faizan Ahmad. "Nanocoating and its application as antimicrobials in the food industry: A review." *International Journal of Biological Macromolecules* 254 (2024): 127906.
29. Kashyap, Prem Lal, Xu Xiang, and Patricia Heiden. "Chitosan nanoparticle based delivery systems for sustainable agriculture." *International Journal of Biological Macromolecules* 77 (2015): 36–51.
30. Ahmed, Farah K., Mousa A. Alghuthaymi, Kamel A. Abd-Elsalam, Mythili Ravichandran, and Anu Kalia. "Nano-based robotic technologies for plant disease diagnosis." In *Nanorobotics and Nanodiagnostics in Integrative Biology and Biomedicine*, pp. 327–359. Cham: Springer International Publishing, 2022.
31. Rajput, Vishnu D., Abhishek Singh, Tatiana Minkina, Sapna Rawat, Saglara Mandzhieva, Svetlana Sushkova, Victoria Shuvaeva et al. "Nano-enabled products: challenges and opportunities for sustainable agriculture." *Plants* 10, no. 12 (2021): 2727.
32. Singh, Kshitij Kumar. "Legal issues in nanotechnology." In *Role of Plant Growth Promoting Microorganisms in Sustainable Agriculture and Nanotechnology*, pp. 107–120. Sawston, UK: Publishing, 2019.

15 Transforming Agriculture Through AR/VR Tools and Technologies

Opportunities and Challenges

Karthigha Mohan, Preethi S L, and Sripriyan K

15.1 INTRODUCTION

Agriculture is one of the most important sectors in the world, and there is a growing need to improve efficiency and productivity in this industry. It has evolved significantly over the years, and modern technology has played a vital role in this evolution. One of the most promising technologies that can help achieve the goal is Augmented Reality (AR) and Virtual Reality (VR). These technologies have been used in several fields, and agriculture is no exception [1]. By providing farmers with real-time data and simulations, AR and VR can improve the overall efficiency and productivity of agricultural practices. They have the potential to transform the way farmers manage their farms and improve sustainability in agriculture. The core features of AR and VR technologies in agriculture include Immersion, Interaction, Integration, Innovation, and Impact. Immersion refers to the ability of AR and VR technologies to create an immersive experience for users [2]. This can provide farmers with a more engaging and effective learning experience, allowing them to acquire new knowledge and skills more quickly. Interaction is another key feature of AR and VR technologies in agriculture. By offering intuitive and natural methods of interaction, these technologies provide farmers with a seamless and user-friendly experience. Integration refers to the ability of AR and VR technologies to integrate with other technologies, such as sensors and machine learning algorithms. By using these technologies together, farmers can access real-time data and insights about their crops, soil, and weather conditions [3]. This can help them to make better decisions and optimize their crop yields. Innovation is another crucial feature of AR and VR technologies in agriculture. By offering new solutions to old problems, these technologies can help farmers to overcome challenges such as crop monitoring, pest and disease detection, and precision farming. This innovation can ultimately lead to more efficient and sustainable agriculture practices. Finally, the impact of AR and VR technologies in agriculture is significant. These technologies have the potential to improve the efficiency and profitability of agriculture, while also promoting sustainability and

DOI: 10.1201/9781003487005-15

reducing waste. Furthermore, by providing farmers with immersive and interactive training experiences, these technologies can help to develop the next generation of skilled farmers. This paper aims to explore the potential applications of AR and VR in agriculture and how they can help improve agricultural practices.

15.2 AUGMENTED REALITY AND VIRTUAL REALITY

15.2.1 Augmented Reality

Augmented Reality (AR) is an exciting technology that enhances our perception of reality by overlaying digital information onto the physical world. It is achieved through the use of sensors, cameras, and displays that allow computer-generated content to be superimposed onto a user's view of the real world. AR applications can be found in a variety of fields, including entertainment, education, advertising, and training. In the entertainment industry, AR is used to create immersive gaming experiences, where players can interact with digital objects and characters in real-world environments [4]. AR is also used in education to create interactive and engaging learning experiences. For example, AR apps can be used to display 3D models of objects that can be explored and interacted with by students. In advertising, AR can be used to create interactive campaigns that engage customers and create a memorable experience. AR can also be used in training to create simulations that provide a safe and controlled environment for learning. Overall, AR is a rapidly evolving technology that has the potential to revolutionize the way we interact with the world around us.

15.2.2 Virtual Reality

Virtual Reality (VR) is a technology that allows users to experience a simulated environment as if they were physically present in that environment. This is achieved through the use of a headset or a set of goggles that display a 3D image or video that responds to the user's movements and actions. In order to interact with the virtual environment, users can use specialized controllers or sensors that track their movements and gestures [5]. This creates a sense of immersion and presence, making the user feel as if they are actually in the virtual world. VR has a wide range of applications, from entertainment and gaming to education and healthcare. In entertainment, VR is used to create immersive experiences such as virtual concerts, theme park rides, and gaming experiences. In education, VR is used to provide interactive and engaging learning experiences, such as virtual field trips to historical sites or science labs. In healthcare, VR is used for pain management and physical therapy, providing patients with immersive experiences that distract them from pain or help them to regain mobility.

15.2.3 Augmented Reality in Agriculture

AR can be used in agriculture to enhance the productivity of farmers. It can provide farmers with real-time information about their crops, soil, and weather conditions. Farmers can use AR to detect different types of pests and diseases in their crops, thereby helping them to take preventive measures early. AR can also be used to create a 3D model of the farm, allowing farmers to visualize their fields and crops in 3D [6]. This can help farmers identify potential problems such as soil erosion or nutrient depletion in specific areas of the farm, and make better decisions about planting, fertilizing, and harvesting. AR can also be used to train new farmers by providing them with an interactive and immersive learning experience. Farmers can learn how to operate machinery, identify different crops and pests, and make decisions based on real-time data. Overall, AR can help farmers make better decisions, reduce waste, and improve the efficiency and sustainability of agriculture.

15.2.4 Virtual Reality in Agriculture

VR has emerged as a promising technology for use in agriculture. VR technology can be used to create immersive simulations of agricultural environments, allowing farmers and researchers to test out different scenarios and make better decisions [7]. It is used in crop management. With VR simulations, farmers can test out different crop management strategies and see the effects on crop yield, soil health, and other factors. This can help them make more informed decisions about how to manage their crops and improve their overall productivity.

Another potential use of VR in agriculture is in training. Farmers and farm workers can use VR simulations to learn about new equipment, safety protocols, and other important aspects of farm management without having to physically be in the field. This can help reduce the risk of accidents and injuries, while also improving overall efficiency and productivity.

15.3 FUTURE DIRECTIONS AND POTENTIAL CHALLENGES FOR THE ADOPTION OF AR/VR IN AGRICULTURE

According to a report by Markets and Markets, the global AR and VR in agriculture market size was valued at USD 0.66 billion in 2019 and is projected to reach USD 2.63 billion by 2025, at a CAGR of 28.4% during the forecast period. This indicates a significant increase in the adoption of AR and VR technologies in agriculture in recent years. Another report by Grand View Research suggests that the use of AR and VR in precision farming is expected to grow rapidly, with a CAGR of 20.6% from 2020 to 2027. This growth is attributed to the increasing demand for real-time data and insights for efficient crop management and the rising adoption of precision agriculture techniques. A survey conducted by the National Farmers' Union (NFU) in the UK found that 30% of farmers are already using some form of precision agriculture technology, with an additional 60% expressing an interest in adopting these technologies in the future. This includes the use of AR and VR technologies for crop monitoring, precision planting, and other applications. Thus, these statistics indicate a growing trend toward the adoption of AR and VR technologies in agriculture. It has the potential to address current challenges faced by farmers and enhance efficiency and productivity. However, there are several potential challenges that may hinder its widespread use. These challenges include the need for technological advancements, reliable and high-speed internet connectivity, integration with existing systems, regulatory barriers, cost, user experience, and acceptance. To overcome these challenges, there is a need for further innovation and development of AR/VR solutions that are affordable, easy to use, and provide tangible benefits to farmers. Regulatory frameworks that address data privacy, intellectual property, and liability issues must also be established to ensure the widespread adoption of AR/VR technology in agriculture. Overall, the success of AR/VR technology in agriculture will depend on overcoming these challenges and continuing to innovate and develop new solutions that meet the specific needs of farmers and the agricultural industry.

15.3.1 Overview of Current Challenges Faced by Farmers and How AR/VR Can Address Them

Farmers face various challenges that affect their productivity and profitability, such as labor shortages, climate change, limited resources, crop diseases, and pests. They are often required to make quick and accurate decisions about crop management, pest control, and resource allocation. However, these decisions are often based on incomplete information and experience, which can result in suboptimal outcomes. For instance, AR/VR technology can enable farmers to monitor crop growth, soil conditions, and water availability in real-time, allowing them to make data-driven decisions about resource allocation. It can also help farmers automate repetitive tasks, such as planting and harvesting, and reduce the need for manual labor.

15.4 TOOLS WIDELY ADDRESSED IN THE FIELD OF AGRICULTURE

In the field of agriculture, various tools and technologies have been widely adopted to improve productivity, efficiency, and sustainability. These tools encompass a range of innovations, including digital solutions, data analytics, precision farming equipment, and more.

15.4.1 FarmVisionAI

FarmVisionAI is an innovative AR platform specifically designed for the agriculture industry. It utilizes cutting-edge computer vision technology to analyze crop health, detect diseases, and optimize pest control. With real-time data and insights, FarmVisionAI empowers farmers to make informed decisions about crop management, leading to improved productivity and sustainability. The platform's features include advanced image recognition algorithms that can identify and classify various crop diseases and pests accurately. It provides farmers with actionable recommendations for targeted treatments, reducing the need for broad-spectrum pesticide applications and minimizing environmental impact. Using FarmVisionAI, farmers can optimize resource allocation, increase crop yields, and reduce production costs. FarmVisionAI has made a significant impact on the agriculture industry by revolutionizing crop monitoring and disease management practices. By leveraging advanced AR technology and computer vision algorithms, the platform has enabled farmers to proactively detect and address crop health issues. The real-time data and insights provided by FarmVisionAI have empowered farmers to make informed decisions about targeted treatments, reducing the reliance on broad-spectrum pesticides and minimizing environmental impact [8].

15.4.2 John Deere See & Spray

John Deere See & Spray is an acclaimed AR tool that has made a significant impact in the agriculture industry. It is specifically designed to address weed management challenges by utilizing advanced computer vision technology. This innovative tool enables farmers to identify and target individual weeds in real-time, resulting in more efficient and sustainable herbicide usage. The key feature of John Deere See & Spray lies in its ability to selectively spray herbicides only on the targeted areas where weeds are present. By using computer vision algorithms, it can differentiate between crops and weeds with high accuracy. This targeted approach minimizes the use of herbicides, reducing costs and environmental impact while promoting more sustainable farming practices. This tool enables farmers to save time and resources by automating the process of weed detection and spraying, improving overall operational efficiency. The AR tool utilizes cameras and sensors mounted on the equipment to analyze the field in real-time. It provides instant feedback and control, allowing farmers to make informed decisions about herbicide application, optimize resources, and ultimately improve crop yields [9].

15.4.3 Plantix

Plantix is an innovative mobile application that has made a significant impact on the agriculture industry. Designed to assist farmers and plant enthusiasts, Plantix combines the power of artificial intelligence (AI) and image recognition technology to provide plant disease diagnostics and crop health monitoring. The app allows users to capture images of their plants, analyze them using the AI algorithms, and receive real-time feedback on potential diseases, nutrient deficiencies, or pest infestations. One of the standout features of Plantix is its vast plant database, which includes a wide range of crops, fruits, and vegetables. This comprehensive library enables accurate identification and diagnosis of plant issues, supporting farmers in making informed decisions regarding pest control strategies and crop management practices. The app offers detailed information on various crops, including growth stages, nutrient requirements, and cultivation techniques. It empowers farmers with timely and accurate information; the app helps them prevent and mitigate crop losses caused by diseases and pests. It has gained widespread usage globally, with millions of farmers, agronomists, and gardening enthusiasts benefiting from its features. The app's user-friendly interface and accessibility have made it popular among smallholder farmers in developing countries, where access to agricultural expertise is limited. By democratizing access to plant diagnostics and knowledge, Plantix is playing a pivotal role in supporting farmers and promoting sustainable agriculture practices worldwide.

15.4.4 Farm AR

Farm AR is an emerging technology that combines augmented reality (AR) with agriculture to revolutionize the way farmers manage their crops and make informed decisions. With its powerful capabilities, Farm AR offers a range of features designed to enhance agricultural practices. By leveraging computer vision, image recognition, and real-time data analysis, farmers can overlay digital information onto their physical environment using AR glasses or mobile devices. The features of Farm AR enable farmers to access vital information about their crops, such as crop health, nutrient levels, soil conditions, and pest infestations. This real-time data empowers farmers to monitor their fields more effectively, identify potential issues early on, and take timely actions to mitigate risks. Through the use of AR, farmers can visualize crop growth patterns, track the progress of their plants, and make informed decisions about irrigation, fertilization, and pest control. It also contributes to reducing chemical usage by enabling targeted treatment of pests and diseases. From large-scale commercial farms to small-scale local growers, this technology benefits farmers of all sizes. It assists in precision agriculture by enabling farmers to identify areas that require attention, such as water stress or nutrient deficiencies, and apply interventions precisely where needed.

15.4.5 AgroScout

AgroScout utilizes image recognition and machine learning algorithms to detect and monitor pests and diseases in crops, helping farmers make informed decisions about crop management. With its comprehensive features and capabilities, AgroScout has made a significant impact on agriculture. One of the key features of AgroScout is its ability to quickly and accurately identify pests and diseases. By capturing images of the crops and processing them through its advanced algorithms, AgroScout can detect and classify various types of pests and diseases, including insects, fungi, and viruses. This enables farmers to take timely action and apply targeted treatments, reducing crop losses and improving overall yields. It also provides real-time monitoring and alerts. Farmers can use the tool to continuously monitor their fields and receive notifications when any abnormalities or potential threats are detected. This allows for proactive intervention, preventing the spread of pests or diseases and minimizing their impact on crop health. The tool collects and analyzes data from multiple sources, including satellite imagery, weather data, and historical records. By integrating this information, AgroScout provides valuable insights into crop health, growth patterns, and environmental factors.

15.5 AR/VR FRAMEWORKS AND HARDWARE DEVICES

15.5.1 Frameworks for App Development

AR/VR tools for app development provide developers with the necessary frameworks, libraries, and software development kits (SDKs) to create augmented reality experiences for various platforms and devices. These tools help streamline the development process and enable developers to build immersive AR applications with ease.

15.5.1.1 Vuforia

Vuforia is an augmented reality (AR) development platform that provides tools and capabilities for creating interactive AR experiences. It offers computer vision technology that enables developers to overlay virtual content on real-world objects and surfaces. It was initially developed by Qualcomm, a mobile technology company, and is now owned by PTC, a global provider of software solutions. It has become one of the leading AR development platforms, widely used by developers and businesses to create AR applications across various industries.

Vuforia offers a range of features and functions for AR development. Its computer vision technology enables the recognition and tracking of real-world objects, such as images, markers, and 3D objects. Developers can use Vuforia to overlay virtual content accurately on these targets. It provides extended tracking capabilities, ensuring a seamless user experience even when the target object is out of view. Smart Terrain functionality allows for the creation of interactive AR experiences in dynamic environments, enabling virtual objects to interact with the physical surroundings and detect obstacles. Cloud recognition is another feature that Vuforia offers, allowing for the recognition of large image databases stored in the cloud. This enhances scalability and flexibility by offloading the recognition process to cloud servers. It also supports major mobile platforms like Android and iOS, as well as popular AR devices such as Microsoft HoloLens. It integrates smoothly with well-known game engines like Unity and Unreal Engine, providing a seamless and efficient development experience.

Vuforia has gained significant traction in the AR development community. Vuforia has been used in over 475,000 projects and powered more than 50,000 applications. It has been adopted by numerous developers, businesses, and educational institutions for a wide range of AR projects. One example of an app related to agriculture developed using Vuforia is the "Plantix" app. Plantix is an AI-powered mobile application that assists farmers and gardeners in identifying plant diseases, pests, and nutrient deficiencies. It utilizes the Vuforia AR platform to overlay informative content and visual markers on the user's device screen.

15.5.1.2 Unity

Unity is a powerful and widely used game engine that has become synonymous with VR app development. It provides developers with a comprehensive set of tools and features to create immersive and interactive virtual reality experiences. Unity is a cross-platform game engine initially developed by Unity Technologies in 2005. It has gained immense popularity due to its versatility, ease of use, and extensive support for multiple platforms, including Windows, macOS, Android, iOS, and many more. Unity allows developers to create interactive 2D and 3D experiences, including virtual reality applications.

Unity offers a wide range of features that make it suitable for VR app development. It provides a visual editor where developers can design and build VR scenes, import assets, and manage project components. It supports scripting in multiple programming languages, allowing developers to write custom scripts for controlling VR interactions and gameplay mechanics. The Unity Asset Store offers a vast marketplace for finding pre-built assets, scripts, and plugins to enhance VR projects. It also includes built-in physics and animation systems for realistic object interactions and character animations. With its cross-platform support, developers can build VR applications that run on popular VR devices like Oculus Rift, HTC Vive, and PlayStation VR. Unity's VR capabilities enable the implementation of various functions in VR applications. Head tracking allows users to explore the VR environment by moving their heads, enhancing immersion. Hand interactions can be implemented using hand tracking or hand-held controllers, enabling users to interact with virtual objects. Unity's audio system supports spatial audio, creating realistic and immersive soundscapes that respond to the user's position and orientation.

Unity has gained significant popularity in the VR app development space. It holds a considerable market share, with over 60% of the top 1,000 mobile games and over 50% of VR and AR experiences developed using Unity. The Unity community is large and active, with millions of registered users worldwide. This community provides valuable resources, tutorials, and support for developers, fostering collaboration and knowledge-sharing. Unity has had a substantial impact on VR applications, driving innovation and creating engaging experiences in various domains. One example of an app related to agriculture developed using Unity is "Farm AR." Farm AR is an augmented reality (AR) application that provides farmers with interactive and informative experiences for crop management and farm planning. The app utilizes Unity's capabilities to create a virtual overlay of data and visuals onto the real-world environment, enhancing the farmer's understanding and decision-making process.

15.5.2 Hardware Devices in AR/VR

In addition to the software development frameworks and platforms mentioned earlier, a wide range of hardware devices can be integrated with the AR/VR development environment, enhancing the immersive and interactive aspects of AR/VR experiences. These devices include mobile devices, cameras, sensors, head-mounted displays (HMDs), and wearables. By integrating these hardware tools, developers can unlock new possibilities and create compelling AR/VR applications. Let's explore some of the hardware tools commonly used.

15.5.2.1 HMDs

Head-mounted displays (HMDs) can be used in agriculture for a range of purposes such as farm machinery and equipment training, AR crop monitoring, remote assistance, and livestock monitoring. HMDs can provide farmers with real-time information and remote assistance to improve safety, reduce downtime, and increase productivity. They can also help farmers identify potential issues quickly and take corrective action before they become serious problems. HMDs equipped with AR and computer vision technology can be used to monitor crop growth, nutrient levels, soil moisture, and livestock health and behavior. However, the cost and complexity of implementing HMDs may be a barrier to adoption for some farmers.

15.5.2.2 Mobile Devices

Mobile devices such as smartphones and tablets are increasingly being used in agriculture for a range of purposes, including field data collection, real-time monitoring of crop growth and livestock health, mapping and GPS navigation in the field, and communication and collaboration between farmers, suppliers, and buyers. These applications have the potential to increase efficiency and productivity in agriculture by providing farmers with real-time information and enabling remote collaboration. However, reliable connectivity and access to the necessary hardware and software may be a challenge in some rural areas.

15.5.2.3 Drone and Robots

Drones and robots are being used in agriculture for crop monitoring, spraying, livestock monitoring, harvesting, and soil analysis. These technologies offer significant potential benefits, including improved efficiency, productivity, and accuracy of data collection. However, there are also challenges such as equipment cost, maintenance, training requirements, and regulations around their use that need to be addressed for wider adoption.

15.5.2.4 Wearables

Wearables are increasingly being used in agriculture to collect and transmit data to a central system for analysis. They include livestock monitoring devices, crop sensors, smart glasses, and GPS tracking devices. Wearable devices are used to monitor the health and behavior of livestock, detect crop stress, provide real-time information on crop health, and track the location of livestock, vehicles, and equipment. The data collected by wearables can be used to optimize irrigation and fertilization practices, detect health issues early, and improve operational efficiency.

15.5.2.5 Sensors and Cameras

Sensors and cameras play a vital role in modern agriculture as they allow farmers to collect real-time data. This data can be analyzed to make informed decisions, improve productivity, and reduce losses. Soil sensors help to optimize irrigation and fertilization, while weather sensors help predict weather patterns for better planting and harvesting decisions. Crop sensors detect signs of stress, optimize pesticide and fertilizer applications, and improve crop yields. Livestock monitoring sensors help to detect health issues early, and cameras can be used for security, surveillance, and monitoring crop growth and damage.

15.6 AR/VR APPLICATIONS IN AGRIBUSINESS

AR/VR applications are making significant strides in revolutionizing the agribusiness sector. From farm management and precision agriculture to supply chain optimization and customer engagement, AR/VR applications are reshaping the way agribusinesses operate in the modern era. Let's delve into the diverse applications and benefits of AR/VR in agribusiness.

15.6.1 Crop Management

AR and VR technologies can help farmers to visualize their crops and fields in three dimensions. This visualization allows farmers to assess the health of their crops, identify areas that need attention, and make informed decisions about irrigation, fertilization, and pest control. This can help to reduce the use of harmful pesticides and minimize the impact on the environment. VR can be used to simulate different weather conditions, allowing farmers to anticipate the effects of climate change on their crops. For example, VR can simulate drought conditions, allowing farmers to test different irrigation strategies to optimize water use and crop yields. It also can be used to train farmers on various aspects of agriculture, including equipment operation, planting techniques, and harvesting practices. It can be used for crop monitoring, which involves the use of sensors and other technologies to monitor crop growth and health. With AR, farmers can monitor their crops in real-time, using sensors and drones, enabling them to identify potential problems early and take corrective action. This can help to reduce crop loss and increase productivity.

15.6.2 Pest and Disease Detection

Pest and disease detection is another critical aspect of crop management in agriculture. The timely detection of pests and diseases can help farmers to take preventive measures, reducing the risk of crop loss and minimizing the use of harmful pesticides. AR and VR can be used to create a digital representation of the crop and its surroundings, allowing farmers to monitor the crop and detect pests and diseases in real-time. For example, AR can be used to overlay digital images on the physical crop, highlighting areas that are affected by pests or diseases. This can help farmers to identify the type of pest or disease and take appropriate action to control it. It can also be used to simulate different scenarios, allowing farmers to test different pest and disease control strategies. This can help to optimize the use of pesticides and reduce the risk of resistance. VR can be used to train farmers on pest and disease identification and control strategies, enabling them to improve their skills and knowledge. It can also be used for remote monitoring of crops. With AR, farmers can monitor their crops from a distance, using drones or other remote sensing technologies. This can help to detect pests and diseases early before they cause significant damage. VR can be used to simulate different weather conditions and their effects on crop growth and health. This can help farmers to anticipate the effects of climate change on their crops and take appropriate action.

15.6.3 Precision Agriculture

Precision agriculture is a modern approach to farming that uses technology to optimize crop production and reduce waste. It involves the use of data-driven techniques to monitor and analyze crops, soil, and weather conditions. It helps farmers to optimize the use of fertilizers, water, and other resources, reducing waste and improving crop yields by creating digital maps of fields and providing farmers with detailed information about soil type, nutrient levels, and moisture content. This can also help farmers to make better decisions about planting and harvesting, reducing the risk of crop failure due to extreme weather events. AR and VR can be used for precision irrigation and fertilization. With AR, farmers can monitor soil moisture levels in real-time, allowing them to adjust irrigation schedules as needed and VR can be used to simulate the effects of different fertilizers on crop growth and health, allowing farmers to optimize fertilizer use and reduce the risk of over-application.

15.6.4 3D Visualization

Three-dimensional (3D) visualization is an emerging technology that has many applications in agriculture. It involves the use of Augmented Reality (AR) and Virtual Reality (VR) technologies to create immersive, interactive, and realistic 3D models of agricultural environments, including fields, crops, and machinery. 3D visualization can help farmers to optimize the layout of their fields and machinery, reducing waste and improving efficiency. VR can be used to train farmers on machinery operation and safety, enabling them to improve their skills and knowledge. With AR, farmers can create interactive, immersive experiences for potential customers, showcasing their products and services in a unique and engaging way. This can help farmers to differentiate themselves from competitors and attract more customers.

15.6.5 Marketing and Sales

AR and VR technologies can also be used for marketing and sales purposes in the agriculture industry. For example, virtual reality can be used to create immersive product demonstrations and farm tours that can help potential customers to better understand the value of products and services. By using AR and VR technologies, companies can provide customers with an interactive experience that allows them to visualize how products work and how they can benefit their farms. Creating 3D models of products and equipment can be used in product catalogs, brochures, and websites. These 3D models provide customers with a more engaging and informative way of learning about products, allowing them to make more informed purchasing decisions. It is used to create virtual showrooms and exhibitions, where customers can interact with products and services without having to travel long distances. This is particularly useful in the agriculture industry, where farms are often located in remote areas, and it can be challenging to attend trade shows and exhibitions. It can be used to create personalized marketing campaigns that target specific customers based on their needs and preferences. By using data analytics and machine learning algorithms, companies can provide customers with targeted and personalized marketing messages that resonate with their interests and needs.

15.7 AGRICULTURE TRENDS IN AR/VR

AR/VR tools are revolutionizing the way farmers and agricultural professionals interact with their crops, livestock, and environments. It is important to explore the latest agriculture trends in AR/VR and understand how these technologies are shaping the future of farming.

15.7.1 Increased Use of Mobile Devices

Mobile devices have become indispensable tool for farmers and agricultural professionals, providing them with easy access to a range of applications and tools. With the rise of Augmented Reality (AR) and Virtual Reality (VR) technologies, mobile devices have become even more critical in agriculture. AR/VR apps allow farmers to visualize their crops and make informed decisions about planting, harvesting, and pest control. These tools are also beneficial in training new farmers and educating consumers about agricultural practices. As technology continues to advance, mobile devices are expected to become even more powerful and affordable, making them more accessible to farmers worldwide. The convenience and portability of mobile devices will continue to play a crucial role in the future of agriculture, transforming the way farmers work and interact with their crops.

15.7.2 Integration with Other Technologies

The integration of AR/VR technologies with other advanced tools has opened up a whole new world of possibilities for industries such as agriculture. One such technology is drones, which are now being equipped with AR/VR technology to capture aerial footage of crops and fields. This enables farmers to analyze their fields in a more comprehensive manner, allowing for more precise and efficient crop management. Integrating it with sensors has enabled farmers to monitor their crops' health and growth more effectively. Sensors can be used to collect data on soil moisture levels, temperature, and nutrient content, which can then be analyzed in real-time using AR/VR technology. This enables farmers to make informed decisions about when to water, fertilize, or harvest their crops. Integration of it with AI has allowed for more advanced data analysis and decision-making. AI algorithms can be trained to recognize patterns in data collected from sensors and drones, enabling farmers to make predictions about crop yields, disease outbreaks, and other important factors. This enables them to plan and execute more effective crop management strategies, leading to increased yields and more efficient use of resources.

15.7.3 Increased Adoption of Immersive Technologies

There has been a noticeable increase in the adoption of immersive technologies across various industries. In agriculture, specifically, the use of VR headsets has gained popularity among farmers and agriculture professionals. By using VR headsets, farmers can experience a realistic simulation of different scenarios that they may encounter in their daily work. This allows them to gain hands-on experience and develop their skills in a more engaging and interactive manner. Furthermore, immersive technologies can also help reduce the risk of accidents and injuries by providing a safe and controlled environment for training.

Apart from training, immersive technologies can also be used to improve decision-making in agriculture. By creating virtual models of fields, farmers can analyze and predict crop growth patterns and make informed decisions about irrigation and fertilization. This can lead to more efficient use of resources and better crop yields. It will be interesting to see how this trend evolves and what other innovative applications emerge in the future.

15.7.4 Emphasis on Sustainability

The pressing need for sustainability has spurred innovative solutions to preserve the planet. One such solution is the integration of AR/VR technologies in agriculture to promote sustainable practices. With AR/VR tools, farmers can now monitor soil and water quality in real-time, reducing waste and optimizing resource management. This technology empowers farmers to make data-driven decisions that improve crop yields while minimizing environmental impact. These tools aid in the identification of potential issues, allowing for timely intervention and mitigating risks. The adoption of sustainable agricultural practices is crucial for the health of our planet, and AR/VR technologies are playing an essential role in advancing this cause.

15.7.5 Development of New Applications

The constant evolution is fostering the development of innovative applications in various fields, including agriculture. As a result, new and exciting applications are emerging to support the growth of the industry. Among these applications are crop simulation tools, virtual marketplaces, and virtual farm tours. These innovative applications enable farmers and agribusinesses to streamline their operations, and promote their products to a broader audience.

Crop simulation tools, virtual marketplaces, virtual farm tours are some of the innovative applications developed for agriculture, revolutionizing the industry and improving the lives of farmers and consumers alike. Consumers are benefiting from these innovations by having access to fresh and locally grown produce, a better understanding of farming practices, and engaging experiences that promote transparency and education in the agriculture industry.

15.7.6 Collaboration

Advanced technologies are enabling stakeholders to collaborate in ways that were previously unimaginable, providing a shared platform for visualizing and analyzing data that can help to optimize agricultural practices and boost yields.

Stakeholders can access real-time data from multiple sources. This data can be analyzed in real-time, providing stakeholders with actionable insights that can be used to make informed decisions about crop management and other key aspects of agricultural operations. It also helps to bridge the gap between farmers and consumers. By providing a virtual window into the world of agriculture, these technologies are enabling consumers to better understand the challenges and opportunities facing the industry, and to make more informed decisions about the food they eat.

15.7.7 Cloud Computing

Farmers are now leveraging the power of cloud-based platforms to store, manage, and analyze their farming data, which is critical for achieving higher yields and better crop quality. By using cloud computing, farmers can access their data from anywhere, at any time, and share it in real-time. This facilitates collaboration and enables quick decision-making, leading to more profitable farming operations.

15.7.8 AI and Machine Learning

AR/VR technologies are being integrated with AI and machine learning algorithms to provide farmers with more accurate and detailed insights about their crops and soil conditions. This information can be used to optimize farming practices and improve crop yields. AR can overlay real-time data onto the farmer's field of vision, while VR can create immersive simulations that allow farmers to explore their fields and experiment with different farming techniques.

15.8 ECONOMIC, POLITICAL AND SOCIAL CONSIDERATIONS SURROUNDING THE USE OF AR/VR IN AGRICULTURE, AND THE IMPACT ON TRADITIONAL FARMING COMMUNITIES

The adoption and integration of AR/VR in agriculture can be influenced by various economic, political, and social factors. Economic factors such as the cost of technology, availability of funding, and profitability can affect the adoption of AR/VR in agriculture. For instance, the high cost of AR/VR technology may limit its adoption among small-scale farmers who may not have the financial resources to invest in such technology. On the other hand, larger agricultural corporations may have the resources to invest in AR/VR technology, which can provide them with a competitive edge in the market. Political factors such as government policies, regulations, and incentives can also influence the adoption of AR/VR in agriculture. For example, governments can incentivize farmers to adopt AR/VR technology by offering tax breaks or subsidies for purchasing and implementing such technology. Additionally, regulations surrounding the use of AR/VR technology, such as privacy laws, can impact its adoption and use. Social factors such as the cultural acceptance of new technology and the perception of its usefulness can also impact the adoption of AR/VR in agriculture. Farmers may be hesitant to adopt new technology if they perceive it to be too complex or time-consuming to use. Also, cultural attitudes towards technology can also impact its adoption. For example, younger generations may be more receptive to the use of AR/VR in agriculture, while older generations may be more resistant to change. Governments, private companies, and other stakeholders need to consider these factors when promoting and implementing AR/VR technology in the agricultural sector. Despite the potential benefits of AR/VR technology, there are still several challenges to be overcome before it can be widely adopted in agriculture. Cost is a significant barrier, as the technology can be expensive to implement and maintain, and also user experience and regulatory barriers need to be addressed to ensure that the technology is user-friendly and complies with relevant regulations.

15.9 CONCLUSION

AR/VR technologies have immense potential to transform the agriculture industry. These innovative tools offer farmers and agricultural professionals a range of applications to improve productivity, sustainability, and decision-making. AR tools can provide real-time data overlays, enabling farmers to monitor crop health, detect pests and diseases, and optimize resource allocation. VR tools offer immersive training experiences, allowing farmers to simulate various scenarios, optimize farm designs, and enhance their skills and knowledge. By leveraging AR/VR in agriculture, farmers can make informed decisions, reduce risks, and ultimately achieve higher yields and more efficient resource utilization. As technology continues to advance, we can expect further advancements in AR/VR tools that will continue to revolutionize the agriculture industry and contribute to its growth and sustainability.

REFERENCES

1. Chai, J. J. K., O'Sullivan, C., Gowen, A. A., Rooney, B., & Xu, J.-L. (2022). Augmented/mixed reality technologies for food: A review. *Trends in Food Science & Technology*, 124, 182–194.
2. Garzon, J., Baldiris, S., Acevedo, J., & Pavon, J. (2020). Augmented reality-based application to foster sustainable agriculture in the context of aquaponics. In *Proceedings – IEEE 20th International Conference on Advanced Learning Technologies, ICALT* (pp. 316–318). https://doi.org/10.1109/ICALT49669.2020.00101
3. Gutiérrez, F., Htun, N. N., Charleer, S., De Croon, R., & Verbert, K. (2019). Designing augmented reality applications for personal health decision-making. In *Proceedings of the 52nd Hawaii International Conference on System Sciences*, Vol. 6 (pp. 1738–1747).
4. Katsaros, A., & Keramopoulos, E. (2017). FarmAR, a farmer's augmented reality application based on semantic web. In *Proceedings of the Computer Networks and Social Media Conference (SEEDA-CECNSM)* (pp. 1–6).
5. Phupattanasilp, P., & Tong, S. R. (2019). Augmented reality in the integrative internet of things (AR-IoT): Application for precision farming. *Sustainability*, 11(9).
6. Todorović, V., Milić, N., & Lazarević, M. (2019). Augmented reality in food production traceability – Use case. In *EUROCON 2019 – 18th International Conference on Smart Technologies* (pp. 1–5).
7. Huuskonen, J., & Oksanen, T. (2018). Soil sampling with drones and augmented reality in precision agriculture. *Computers and Electronics in Agriculture*, 154, 25–35.

8. Liu, M., Li, X., Lei, X., & Wu, S. (2013). Research of mobile augmented reality technology applied in agriculture. In *Proceedings of the 2013 International Conference on Advanced Computer Science and Electronics Information* (pp. 311–315). Beijing, China.
9. Zhao, Z., Yang, W., & Chinthammit, W. (2017). A new approach to utilize augmented reality on precision livestock farming. In *Proceedings of the International Conference on Artificial Reality and Telexistence Eurographics Symposium on Virtual Environments* (pp. 22–24). Adelaide, Australia.

Index

FSC
www.fsc.org
MIX
Papier | Fördert
gute Waldnutzung
FSC® C083411

Kolibri 360 GmbH
Ferdinand-Jühlke-Straße 7
99095 Erfurt, Deutschland
produktsicherheit@kolibri360.de